International Conference on Robotics and Mechantronics (ICRoM 2017)

IOP Conference Series: Materials Science and Engineering
Volume 320

Hong Kong, China
12 – 14 December 2017

ISBN: 978-1-5108-5927-2
ISSN: 1757-8981

Printed from e-media with permission by:

Curran Associates, Inc.
57 Morehouse Lane
Red Hook, NY 12571

proceedings.com

Some format issues inherent in the e-media version may also appear in this print version.

Printed by Curran Associates, Inc. (2018)

For permission requests, please contact the Institute of Physics
at the address below.

Institute of Physics
Dirac House, Temple Back
Bristol BS1 6BE UK

Phone: 44 1 17 929 7481
Fax: 44 1 17 920 0979

techtracking@iop.org

Additional copies of this publication are available from:

Curran Associates, Inc.
57 Morehouse Lane
Red Hook, NY 12571 USA
Phone: 845-758-0400
Fax: 845-758-2633
Email: curran@proceedings.com
Web: www.proceedings.com

Table of contents

Volume 320

International Conference on Robotics and Mechantronics (ICRoM 2017)
12–14 December 2017, Hong Kong

Accepted papers received: 16 February 2018
Published online: 5 March 2018

Preface

Papers

Robot System Design and Motion Modeling

IOP Conf. Series: Materials Science and Engineering **320** (2018) 011001 doi:10.1088/1757-899X/320/1/011001

Preface

It is our great pleasure to welcome you to 2017 the International Conference on Robotics and Mechantronics (ICRoM 2017), which will be held in Hong Kong Polytechnic University during December 12-14, 2017.

The objective of ICRoM 2017 has always been bringing together a large group of researchers, scientists, academics, and engineers in this related area from all over the world, who share much more than technical interests, among other things, culture and history.

This proceedings of ICRoM 2017 presents papers submitted to the conference from universities, research institutes and industries. The papers selected depended on their quality and their relevancy to the conference. The volume tends to present to the readers the recent advances in the field of mechanical engineering and robot development technology, electronic information engineering and application, control theory and engineering etc.

We would like to thank all the authors who have contributed to this volume and also to reviewers, speakers, chairpersons, sponsors and all the conference participants for their hard work and support to ICRoM 2017.

ICRoM 2017 Organizing Committee
December 12, 2017

IOP Conf. Series: Materials Science and Engineering **320** (2018) 011002 doi:10.1088/1757-899X/320/1/011002

Conference Committee

General Chair

Prof. Ka-wai Eric Cheng, The Hong Kong Polytechnic University, Hong Kong

Technical Program Chairs

Prof. Meng Joo Er, Nanyang Technological University, Singapore

Prof. W.Z. Cui, Chongqing University, China

International Management Committee

Prof. Chi-Fang Huang, Institute of Communication of Engineering, Tatung University, Taipei, Taiwan

Roy Kalawsky, Loughborough University, UK

Akalanka Weerasooriya, Volunteer, IEEE Sri Lanka section

Darmawan Sutanto, University of Wollongong, Australia

Pablo Arboleya, Universidad de Oviedo, Spain

Charles R. Sullivan, Dartmouth College, US

Fei Gao, University of Technology of Belfort-Montbeliard, France

Lei Dong, Beijing Institute of Techonology, China

Prof. Divakar B. P. , Reva University, India

Dr. Man-Chung WONG, University of Macau, Macau

Peer review statement

All papers published in this volume of *IOP Conference Series: Materials Science and Engineering* have been peer reviewed through processes administered by the proceedings Editors. Reviews were conducted by expert referees to the professional and scientific standards expected of a proceedings journal published by IOP Publishing.

IOP Conf. Series: Materials Science and Engineering **320** (2018) 012001 doi:10.1088/1757-899X/320/1/012001

Kinematics Simulation Analysis of Packaging Robot with Joint Clearance

Y W Zhang[1,2], W J Meng[2], L Q Wang[2], G H Cui[2]
1 Automotive Engineering College, Shanghai University of Engineering Science, Shanghai 201620, China
2 Mechanical and Equipment Engineering College, Hebei University of Engineering, Handan 056038, China
Abstract. Considering the influence of joint clearance on the motion error, repeated positioning accuracy and overall position of the machine, this paper presents simulation analysis of a packaging robot — 2 degrees of freedom(DOF) planar parallel robot based on the characteristics of high precision and fast speed of packaging equipment. The motion constraint equation of the mechanism is established, and the analysis and simulation of the motion error are carried out in the case of turning the revolute clearance. The simulation results show that the size of the joint clearance will affect the movement accuracy and packaging efficiency of the packaging robot. The analysis provides a reference point of view for the packaging equipment design and selection criteria and has a great significance on the packaging industry automation.

1. Introduction

Packaging industry automation is the inevitable result of the development of science and technology. The movement of packaging equipment, the core of the entire production line part, directly affects the operating efficiency. Now a great many of parallel robot play a role of a box robot in the production[1] as the result of high stiffness, high precision and dynamic performance characteristics of the parallel mechanism. However, because of the joint clearance and other factors, the advantages of the mechanism cannot fully play out[2-5]. Therefore, it is very important to choose the packaging robot in the case of kinematics simulation analysis with the joint clearance. Some progress[6-11] has been made in the study of the impact of motion pair clearance on the performance of the mechanism. The earliest study was the revolute model of the plane mechanism [12]. Biswas [13] introduced a planar mobile clearance model in the literature. Wang [14] first established the space rotation clearance model.

Based on the previous research results, this paper analyzes the motion error of a packaging robot - 2-DOF planar parallel robot with the joint clearance.

2. Working principle of packaging robots

IOP Conf. Series: Materials Science and Engineering **320** (2018) 012001 doi:10.1088/1757-899X/320/1/012001

2-DOF planar parallel robot is mainly composed of the frame, moving platform and two branches connecting the frame with moving platform, as shown in figure 1. The left branch consists of the active arm and the slave arm, and the slave arm is composed of two symmetrical single rods, which increases the stiffness and carrying capacity of the branch. The right branch is composed of the active arm, the active arm link, the hinge plate, the slave arm and the slave arm link, the link connecting the active arm with the active arm and the link connecting the slave arm with the slave arm form two parallelogram structures by hinge plates [15]. Two sets of parallelogram structure not only limit the rotation of the end, to achieve the end position and attitude of the decoupling. At the same time, the parallelogram structure can also improve the overall stiffness of the robot end in the vertical direction[16]. Both the active link, the slave link and both ends of driven arm are hingedly connected with a smaller angle to reduce assembly errors causing by manufacturing errors. The whole parallel robot all use revolute pair to achieve a single plane translating along x, y direction, to improve the speed and the acceleration of the moving platform and to satisfy a fast, stable packing operations requirements.

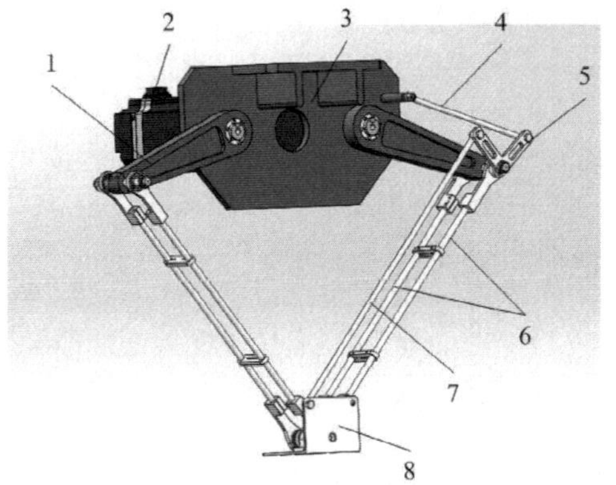

1-Active arm; 2-Servo motor; 3-Frame; 4-Active arm link; 5-Hinge plate;
6-Slave arm; 7-Slave arm link; 8-Moving platform

Figure 1. 2-DOF planar parallel robot.

3. Kinematic analysis of mechanisms

Taking the 2-DOF planar parallel robot as the object of study, the mechanism can be simplified into a symmetrical five link articulated mechanism in the case of the kinematic analysis of the parallel mechanism. The schematic diagram of the mechanism is shown in figure 2:

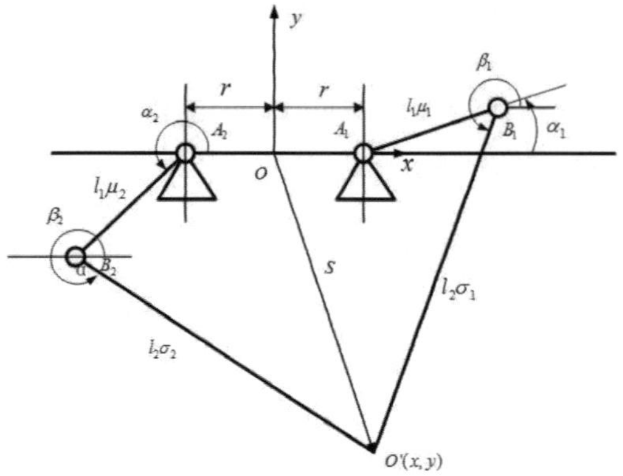

Figure 2. Structure diagram of 2-DOF planar parallel robot.

The symmetry center point of the two active arms is the origin, and the reference coordinate system $O-xy$ is shown in figure 2. $2r$ is the distance between the rotating shaft of the active arm of the frame; l_1 is the robot's active arm length; μ_2、μ_1 are the unit vector of the left and right active arm; l_2 is the slave arm length; σ_2 and σ_1 are the unit direction vectors of left and right driven arms; $O'(x,y)$ is the moving platform position vector; s is the position vector of O' with respect to O. The joint angle of the robot's active arm is α_1 and α_2, and the rotation angle of the driven arm joint is β_1 and β_2.

As shown in figure 2, the position of the mechanism is solved with respect to the known active arm angle $\alpha_i\,(i=1,2)$, and the position vector $O'(x,y)$ of the moving platform is solved. According to the mechanism geometry [17], the following equations are constructed in the reference coordinate system.

$$\begin{cases}-r+l_1\cos\alpha_2+l_2\cos\beta_2-l_2\cos\beta_1-l_1\cos\alpha_1+r=0\\ l_1\sin\alpha_2+l_2\sin\beta_2-l_2\sin\beta_1+l_1\sin\alpha_1=0\end{cases}\tag{1}$$

The moving platform position vector reference point m can be expressed as:

$$\begin{cases}x=-r+l_1\cos\alpha_2+l_2\cos\beta_2\\ y=l_1\sin\alpha_2+l_2\sin\beta_2\end{cases}\tag{2}$$

or

$$\begin{cases}x=r+l_1\cos\alpha_1+l_2\cos\beta_1\\ y=l_1\sin\alpha_1+l_2\sin\beta_1\end{cases}\tag{3}$$

4. Kinematic error analysis considering the revolute clearance

For packaging robot, servo motor and the driving arm connection clearance is not obvious in the mechanical structure, and between the driving arm and the driven arm, a driven arm and moving platform with deep groove ball bearings, there is a clearance of a certain size. Therefore, the clearance between B_1, B_2 and O' is considered in the analysis.

4.1. model of revolute clearance

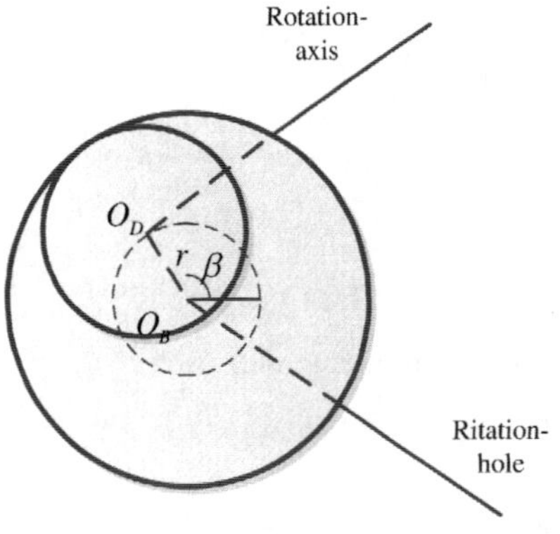

(a) Clearance model

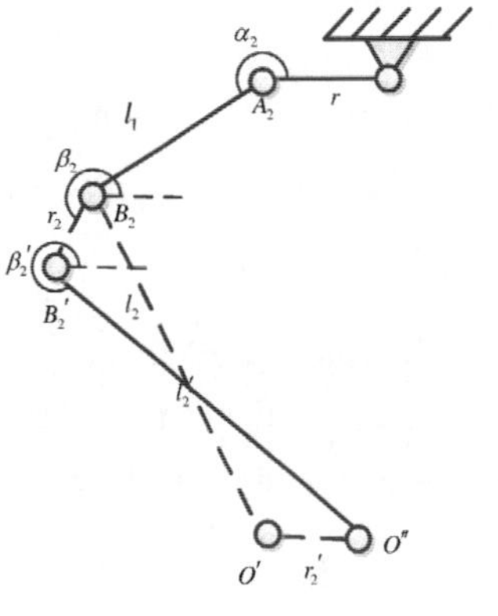

(b) Equivalent length of slave arm

Figure 3. Rotation clearance model.

Model of revolute clearance uses two-state motion model [6], that is, only consider contact and separation of two states for the revolute clearance, and the revolute clearance is equivalent to a non-quality virtual link, as shown in figure 3(a). Assuming that the radius and the axis radius of the revolute pair are R_B and R_D, respectively, the size of the motion pair clearance c is

$$c = R_B - R_D \qquad (4)$$

Due to the existence of the revolute clearance, the axis of the revolute pair can move freely in the hole, and the virtual massless link can be described by the virtual massless link length N and the corner M. Their range of value is

$$r \in [0, c]; \beta \in [0, 2\pi] \qquad (5)$$

Considering the existence of virtual massless link, the length of the follower arm changes l_2' from l_2, and the angle changes β_2' from β_2, the value of virtual massless link l_2' has a range of

$$l_2 - (r_2 + r_2') \le l_2' \le l_2 + (r_2 + r_2') \qquad (6)$$

The link length error caused by the revolute clearance is

$$\Delta l_2 \in [-(r_2 + r_2'), (r_2 + r_2')] \square \ l_2 \qquad (7)$$

The length error of the link caused by the clearance is much smaller than the length of the link.

4.2. Model for revolute error transfer

The equation (2) and the equation (3) are integrated only in case that the clearance error of the revolute pair is taken into account.

$$\begin{cases} x = jr + l_1 \cos\alpha_i + l_2 \cos\beta_i \\ y = l_1 \sin\alpha_i + l_2 \sin\beta_i \end{cases} \qquad \left(\begin{matrix} i = 1, j = 1 \\ i = 2, j = -1 \end{matrix} \right) \qquad (8)$$

Partial derivation for l_i, β_i and $O'(x, y)$ can derive

$$\begin{cases} \Delta x = \Delta l_2 \cos\alpha_i - l_2 \sin\beta_i \Delta\beta_i \\ \Delta y = \Delta l_2 \sin\alpha_i + l_2 \cos\beta_i \Delta\beta_i \end{cases} \qquad i = 1, 2 \qquad (9)$$

Write the above formula as a matrix

$$\begin{bmatrix} 1 & 0 & -l_2 \sin\beta_1 & 0 \\ 0 & 1 & l_2 \cos\beta_1 & 0 \\ 1 & 0 & 0 & l_2 \sin\beta_2 \\ 0 & 1 & 0 & l_2 \sin\beta_2 \end{bmatrix} \begin{bmatrix} \Delta x \\ \Delta y \\ \Delta\beta_1 \\ \Delta\beta_2 \end{bmatrix} = \begin{bmatrix} \beta_1 & 0 \\ \beta_1 & 0 \\ 0 & \beta_2 \\ 0 & \beta_2 \end{bmatrix} \begin{bmatrix} \Delta l_2 \\ \Delta l_2 \end{bmatrix} \qquad (10)$$

Where Δl_2 is the virtual link length error caused by the revolute clearance. Through the above formula, the error from packaging robot and arm angle can be easily obtained.

4.3. Analysis on error simulation results

According to the contact model of the material parameters, the clearance size of the pair is determined to be 0.1mm, which is mainly for the error analysis of the moving platform of the box robot to the position of the grab.

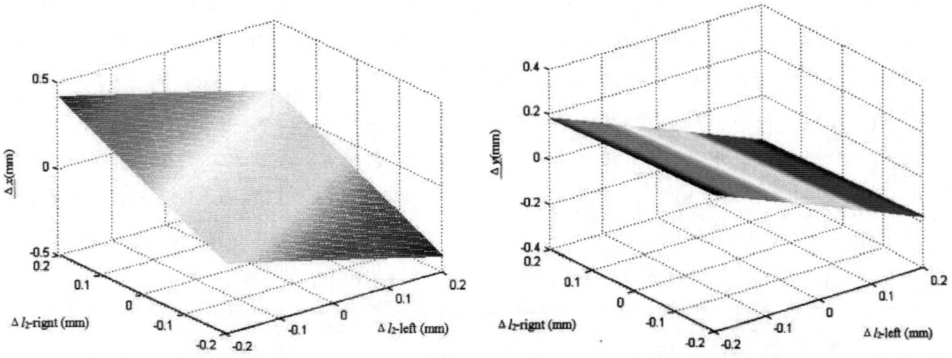

(a) The moving platform error along the *x* direction (b) The moving platform error along the y direction

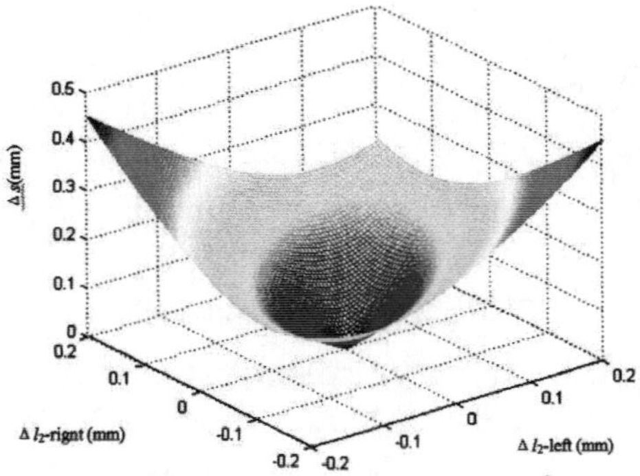

(c) Total error of moving platform

Figure 4. Position error of moving platform.

It can be seen from figure 4 that the errors caused by the revolute clearance are symmetrical about the axis $y = x$, and the maximum error occurs at the positions where errors of the left and right slave arm is $\Delta l_2(-0.2,0.2)$ and $(0.2,-0.2)$, where the angular variation of the left and right arms is $(-5.0142^{e^{-4}},5.0142^{e^{-4}})$ rad and the direction error of the moving platform is $(-0.4111,0.4111)$ mm. The variation range of the moving direction is $(-0.2049,0.2049)$ mm; the total error range of the moving platform is $(0,0.4482)$ mm. The range on position error of robot movement platform is strictly control in 0.5 mm to meet the packing operation requirements.

5. Simulation analysis of packaging robots with the revolute clearance

After analyzing the kinematic error of the packaging robot, a model with a clearance of 0 mm, 0.1 mm, and 0.2 mm was established by Solidworks. In the Adams, the contact force function IMPACT is used to add the parameters such as stiffness, material damping, stiffness index, penetration depth, static friction speed, dynamic friction speed, static friction coefficient, dynamic friction coefficient and collision recovery coefficient of two contact parts. The influence of the motion error caused by the rotation clearance on the pose of the whole body is simulated.

The simulation results show that the curves of the driven arm angle and the moving platform position are shown in figure 5. From figure 5(a) and (b), it can be seen that the curve of the rotation angle from the slave arm is smooth and round when there is no clearance, and when the rotation clearance is 0.1mm and 0.2mm,there is a small floating up and down, but the overall basic coincidence for the

IOP Conf. Series: Materials Science and Engineering **320** (2018) 012001 doi:10.1088/1757-899X/320/1/012001

curve of the slave arm angle is based on the clearance-less curve; From figure 5(c), (d) and (e), it can be seen that when the clearance value is 0.1 mm and 0.2 mm, the moving plate along the x direction, the y direction and the position curve are basically coincident with the clearance-less curve, But there is a slight fluctuation. However, considering the above-mentioned variation curve, it can be concluded that the change of the pose of the whole body is increasing with the increase of the rotation clearance, and the moving platform is more obvious away from the ideal position. The size of the rotation clearance will affect the motion precision and packing efficiency. So the size of the clearance for hole and shaft conjugation should be strictly controlled in the machining process.

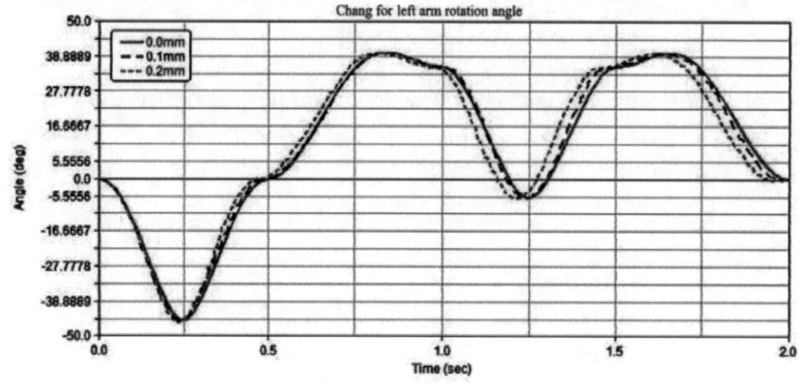

(a) Chang for left arm rotation angle

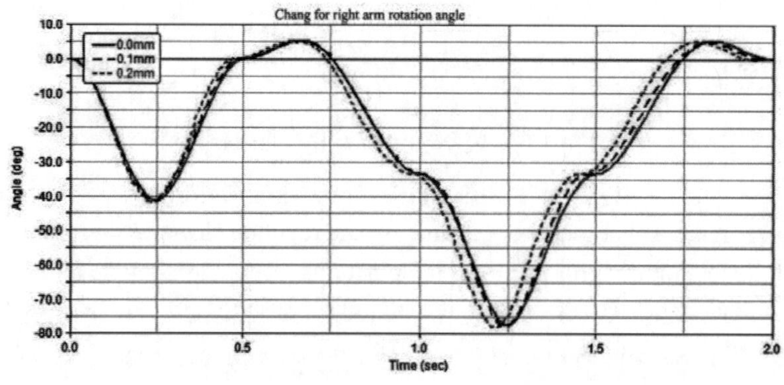

(b) Chang for right arm rotation angle

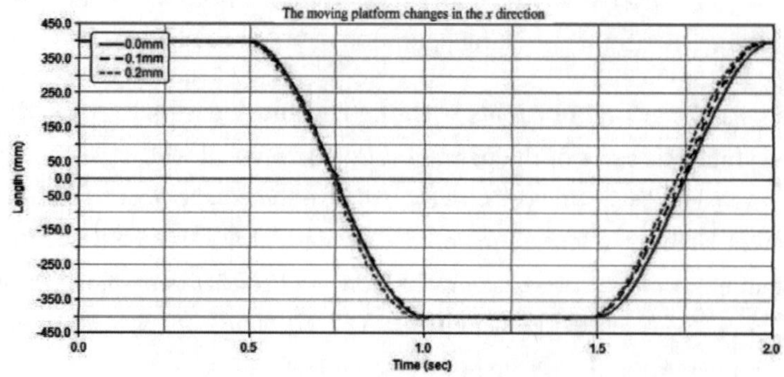

(c) The moving platform changes in the x direction

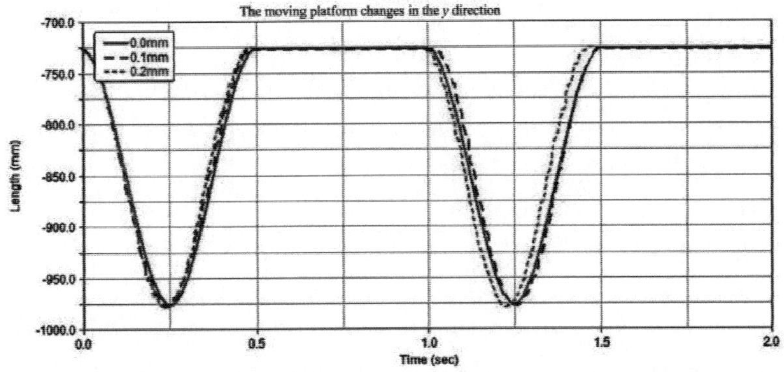

(d) The moving platform changes in the y direction

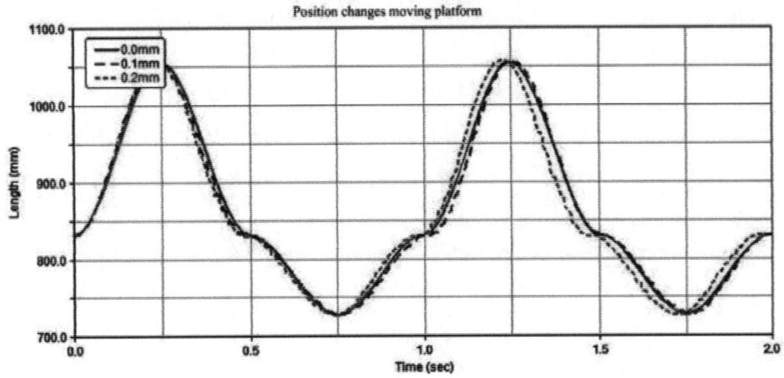

(e) Position changes moving platform

Figure 5. Curve of different clearance for packing robot.

6. Conclusion

Through the theoretical research and simulation analysis of the mechanism on a packaging robot, the following conclusions are obtained:

1) In this mechanism, verify that the mechanism meets the packing requirements by considering the motion error analysis of the rotation clearance, Which provides a reference direction for improving the precision of the selected and designed packing equipment.

2) Through the simulation analysis of the whole robot, it is found that the motion accuracy becomes lower as the rotation clearance increases. Therefore, the size of the clearance for hole and shaft conjugation should be strictly controlled in the machining process.

The author only takes into account the error of the box robot caused by rotating the auxiliary gap, does not consider the error caused by the geometric size and assembly precision, and does not compensate for the error caused, which will become the further research direction.

7. Acknowledgments

This research was supported by the Natural Science Foundation of Hebei Province (grant no. E2015402130)

8. Reference

[1] Yi L, 2015 High-speed Parallel Robot into the Field of Packaging Automation *China Packaging Industry* vol 19, pp 24-25.

[2] Chen F, Li X L, She X, *etal.* 2017 Reliability Design of Planar Path Mechanism with Joint Clearance *Mechanical Science and Technology for Aerospace Engineering* vol 36, pp 863-868.

[3] Yu Y Q and Tian H 2015 Error and Compensation of Parallel Robot with Joint Clearances *Optics and Precision Engineering* vol 23, pp 1331-1340.

[4] Zhang X C, 2015. Research on the Performance of Planar 3DOF Parallel Mechanisms with Joint Clearances. Guangzhou: South China University of Technology.

[5] Ding J, Zhao Y, Wu H T, *etal.* 2014 Dynamic Characteristics of Mechanisms with Multi-clearance Joints *Modular Machine Tool and Automatic Manufacturing Technique* vol 5, pp 9-13.

[6] Huang H J, Zhang C, Jin X, *etal.* 2016 Simulation for Dynamic Balancing of a Scroll Compressor with Joint Clearance *Journal of Vibration and Shock* vol 35, pp 125-130.

[7] Si Z J, 2015. Dynamic Study for Working Mechanism of Crusher with Multi-joint Clearance. Inner Mongolia:Inner Mongolia University of Science and Technology.

[8] Zhi C J, Wang S M and Sun Y T 2014 Effect of Kinematic Pairs Clearance on Dynamic Characteristics of Multi-linkage Lock Mechanisms *Journal of Harbin Institute of Technology* vol 46, pp 102-106.

[9] Zhang Z L, Wu S J, Zhao W Q, *etal.* 2014 Dynamic Characteristics of High-speed Multi-link Transmission Mechanisms with Clearance *Journal of Vibration and Shock* vol 33, pp 66-71.

[10] Wang X, 2014. Dynamics Simulation Research fo Crusher with Multi-joint Clearance.Inner Mongolia:Inner Mongolia University of Science and Technology.

[11] Wang G Q and Liu H Z 2001 Dynamic Response of Crank-rod of an IC Engine with Clearance *Transactions of the Chinese Society for Agricultural Machinery* vol 6, pp 5-7.

[12] Earles S W E and Wu C L S 1972 *Conference on Mechanisms* (England: London) pp 83-89.

[13] Biswas A and Kinzel G L 1998 *Proceeding of 1998 ASME Design Engineering Technical Conference* (USA: Atlanta)

[14] Wang H H S and Roth B 1989 Position Errors Due to Clearances in Journal Bearings *ASME Journal of Mechanical Design* vol 111, pp 315-320.

[15] Du H T 2009 *Chinese Patent*: 200820106420.8

[16] Yan Z G, Wang L Q, Sun Z H, *etal.* 2009 Study of High-speed Auto-sorting System for Food Production *Packaging Engineering* vol 30, pp 16-18.

[17] Zhang C Y 2015. Dynamics Simulation of Space Robot Arm's Clearance and Friction. Qinghuangdao: YanShan University.

Design and Dynamic Modeling of Flexible Rehabilitation Mechanical Glove

M X Lin[1], G Y Ma[1], F Q Liu[1], Q S Sun[1], A Q Song[2]

[1] School of Mechanical Engineering, Shandong University, Jinan, 250061, China
[2] Shandong Provincial Western Hospital, Jinan, 250022, China

Abstract. Rehabilitation gloves are equipment that helps rehabilitation doctors perform finger rehabilitation training, which can greatly reduce the labour intensity of rehabilitation doctors and make more people receive finger rehabilitation training. In the light of the defects of the existing rehabilitation gloves such as complicated structure and stiff movement, a rehabilitation mechanical glove is designed, which provides driving force by using the air cylinder and adopts a rope-spring mechanism to ensure the flexibility of the movement. In order to fit the size of different hands, the bandage ring which can adjust size is used to make the mechanism fixed. In the interest of solve the complex problem of dynamic equation, dynamic simulation is carried out by using Adams to obtain the motion curve, which is easy to optimize the structure of ring position.

1. Introduction

1.1. Background
With the aging of the population, there are a large number of patients with hemiplegia, whose motor function is damaged, which seriously affects the normal life. Rehabilitation medicine has proved that scientific rehabilitation training plays a very important role in the recovery of motor function. At present, these patients rely on rehabilitation physicians to help patients complete rehabilitation training. However, due to the problems of personnel and expenses, many patients can not receive effective rehabilitation training. Rehabilitation robot can relieve the heavy work of rehabilitation therapists and provide more scientific rehabilitation for patients. It is an ideal rehabilitation training tool. Because finger is the most flexible and fragile part of human joints, the rehabilitation robot for finger is the focus of research on motion rehabilitation robot.

1.2. Research status
As early as 2005, Wege et al. (Technical University of Berlin) designed an exoskeleton rehabilitation manipulator (Figure 1) for rehabilitation training [1-2]. Harbin Institute of Technology developed an index finger rehabilitation machine (Figure 2), and added force sensors to detect motion state [3-4].
M.Fontana, F. Salsedo and M.Bergamasco designed a finger rehabilitation treatment machine of the index finger and thumb hand (Figure 3). Because of the coupling between the fingers, only 3 degrees of freedom are used for motion control of the finger. In other words, the fingertip can be driven without joints [5]. Then, Santa Ana college pizza has developed is applied to the index finger

exoskeleton robot HANDEXOS (Figure 4). HANDEXOS has 6 degrees of freedom, which can realize the extension, flexion and abduction/adductionof the index finger [6].

Figure 1. A Hand Exoskeleton for Rehabilitation by the Technical University of Berlin.

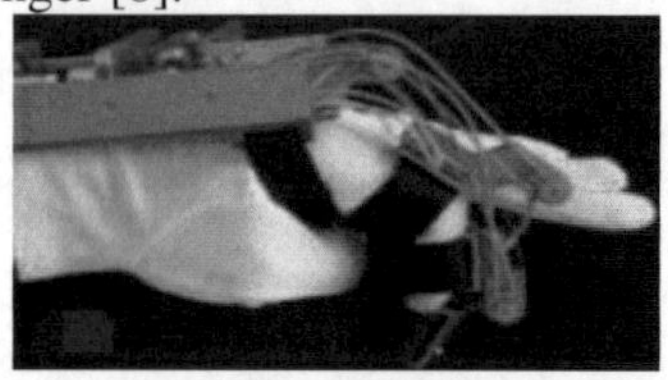

Figure 2. An index finger rehabilitation machine by Harbin Institute of Technology.

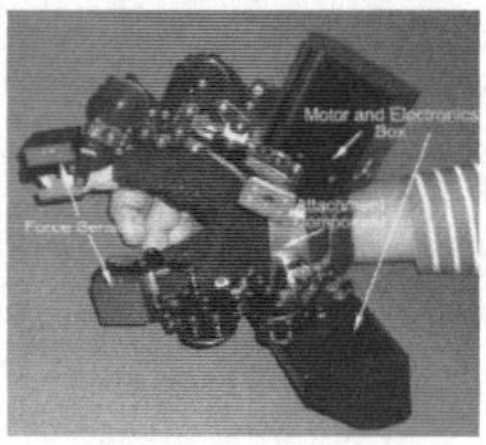

Figure 3. Index and thumb finger rehabilitation robot hand.

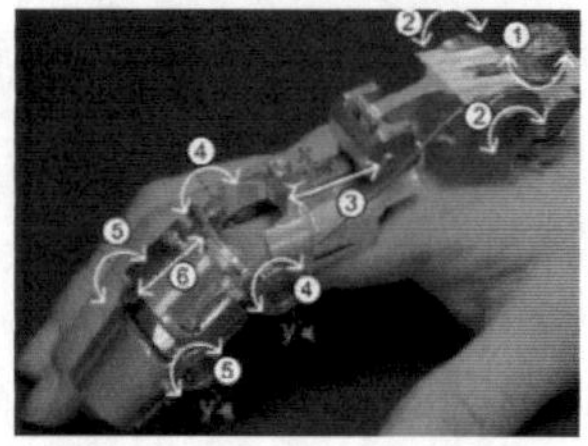

Figure 4. Index finger rehabilitation robot – HANDEXOS.

Because the rigid movement is not conducive to the rehabilitation effect, Mohd Nor Azmi Bin Ab Patar et al developed the exoskeleton rehabilitation robot based on pneumatic muscle (figure 5). Then the finger movement posture is analyzed, which can improve the effect of rehabilitation movement [7].

Figure 5. Pneumatic muscle driven exoskeleton rehabilitation gloves.

1.3. The main content of this paper

The rehabilitation robot has made great progress in recent years, but it has many problems, such as huge structure, complex movement, lack of flexibility and lack of comfort. In this paper, the structure of rehabilitation mechanical glove is designed by using the soft material. The underactuated structure by rope-spring is adopted. Then the mathematical model of dynamics is established, and the simulation of ADAMS is made.

2. Structural design

2.1. Mechanical structure design

From the above analysis, it can be seen that the rehabilitation gloves not only need to realize the stretching and flexion movement of fingers, but also keep the concise structure and flexible movement design. For this reason, pneumatic method is adopted in this paper, which can reduce the volume of driving structure and have the flexibility of pneumatic drive. The rehabilitation gloves are shown in figure 6 and figure 7. Each finger knuckle has a bandage fastened to form a ring. The transmission mechanism adopts combination the rope with the rubber band. The rope is positioned on the palm of the hand and is positioned at the middle of the finger and the ring. One end of the rope is fixed on the ring of the outermost ring, and the other end is fixed with the air cylinder output lever. When the air

cylinder output lever retracts, the rope pulls the finger to complete the flexion movement (Figure 8). The rubber band is placed on the back of the hand, and one end of the rubber band is fixed on the outermost ring, while the other end is fixed on the ring of the hand back which is immobilized with the hand. When the air cylinder output lever extends, the pull from rope has been lost, meanwhile the elastic force of rubber band guides finger flexion.
Because the finger belongs to the weak part of the body, generally the finger rehabilitation training only needs dozens of Newton force. For this reason, the CJ2B10-40 cylinder of SMC company is selected as the drive. Its inner diameter is 10mm with the 0.7MPa maximum pressure, therefore the maximum output force is $F=\pi*(d_{in}/2)^2*P_{\max}=\pi*(10mm/2)^2*0.7MPa=55N$, which could provide effective output.

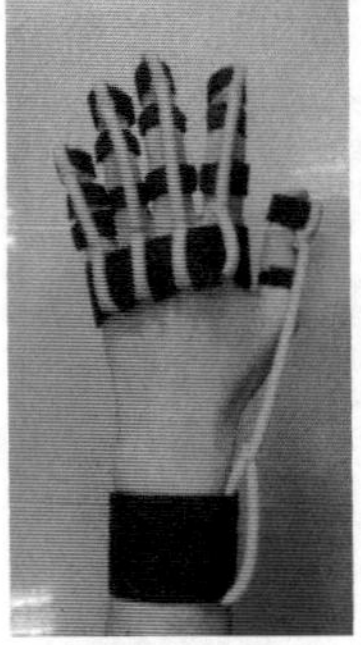

Figure 6. The dorsum of the rehabilitation glove.

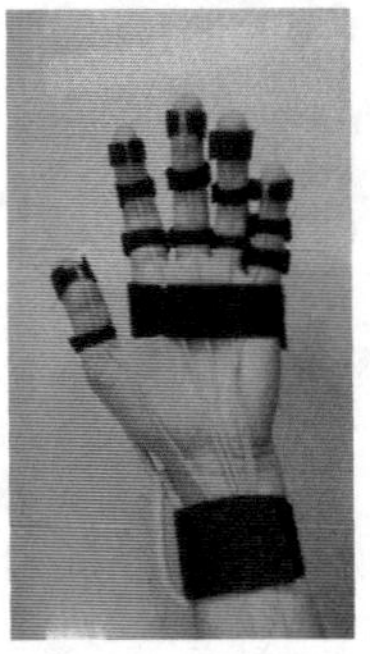

Figure 7. The palm of the rehabilitation glove.

Figure 8. Flexion status of the Rehabilitation glove.

2.2. Control system design
The design of the control system mainly includes two parts: electrical system and pneumatic system. Because of its portable characteristics, the rehabilitation gloves are powered by a lithium battery. In accordance with the battery, the small 12V-DC air compressor is selected to provide air pressure. The current of the selected air compressor is 8A, and the rehabilitation training takes about half an hour. For this reason, battery capacity is required to be greater than 4000mAh, and the battery of 6800mAh is selected in this paper. The capacity of each cylinder is $V_{each}=\pi*(d_{in}/2)^2*l_{cylinder}$ $=\pi*(10mm/2)^2*40mm=3.14*10^{-3}L$. Considering of the trachea gas, $0.2L$ gasholder is determined to ensure the safe of this system. The two-position five-way solenoid valve 4V210-08-12V is selected to complete the two actions, expansion and contraction, of the cylinder. At the same time, the five-floors bus board is used to increase the overall system compactness. The STM32 with 5 relays is selected as the master control circuit board. The details of the control system are shown in Figure 9.

3. Dynamics simulation

Because the theoretical formula is difficult to solve, the dynamic simulation of finger with rehabilitation gloves is carried out by using virtual prototype simulation software Adams in order to analyse the relationship between driving force and motion of each joint intuitively. In this paper, the rehabilitation gloves with index finger are simulated as an example. The model is composed of three parts: the human index finger model, the flexible rope and the flexible spring.
The human index finger model resembles a series of three connecting rods. According to the length and weight of the human index finger, the corresponding simplified model is established in ADAMS. Subsequently, according to the relationship of the motion among the joints, the corresponding constraint is added to the index finger model. The most widely used modeling method is discrete method for the establishment of flexible rope, that is, flexible rope is discretized into finite elements of cylinders, and bushing connections between cylinders are used. As this method can effectively

simulate the actual stress state of flexible rope, it is used to build flexible ropes in this paper. The "Create Spring Resistance" command is used in Adams software for the establishment of flexible springs. The index finger model in Adams is shown in Figure 10.

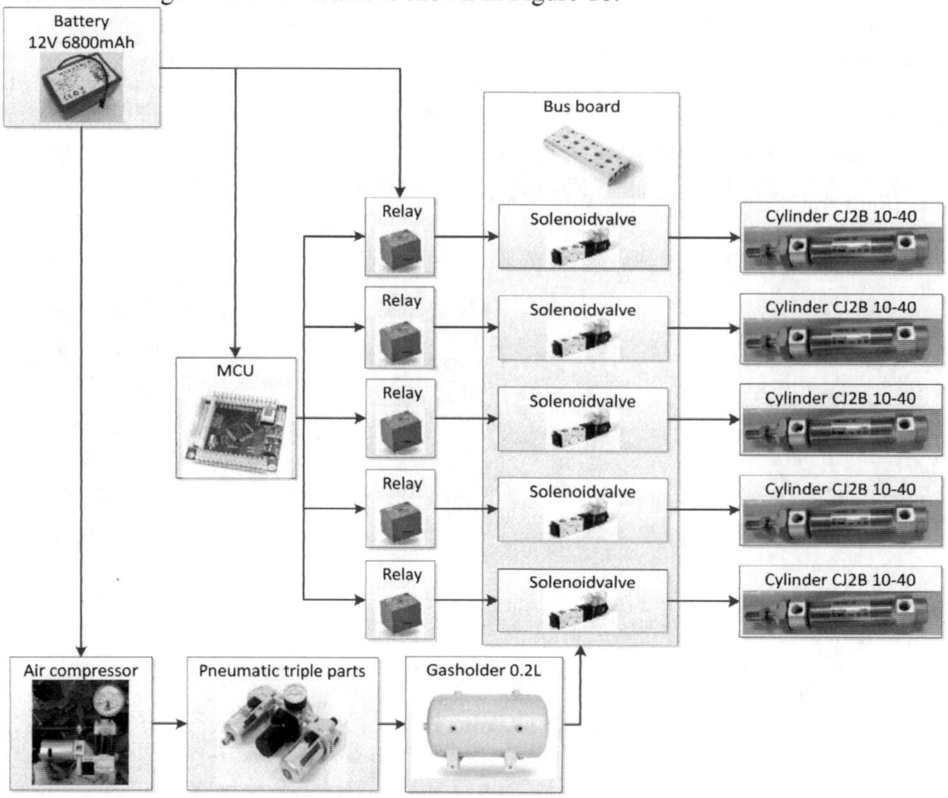

Figure 9. Rehabilitation glove control system.

When the forefinger is in the extended state, the muscle of distal and middle knuckles and the muscle of middle and proximal knuckles are in the state of tension. And springs are added between the two joints to simulate the pull of the muscles, which ensures that the movement of the model is more consistent with the real state of the movement of the finger.

After the model is created, various constraints and forces are added to the model. The joints of the finger are connected by rotating pairs; the lower part of the rope and the knuckles connected by "contact force"; the constraint between the upper part of the rope and the knuckle is defined as fixed connection. In actual rehabilitation training, the driving force of the fingers is about 4-5*N*, therefore the driving force is set to 4.5*N*. The simulation time is set to 0.45*s*, and the dynamics simulation is shown in figure 11. And the post-processing module of Adams software is used to output the rotation angles of the distal joint, middle joint and proximal joint under the action of *F*, which is shown in figure 12.

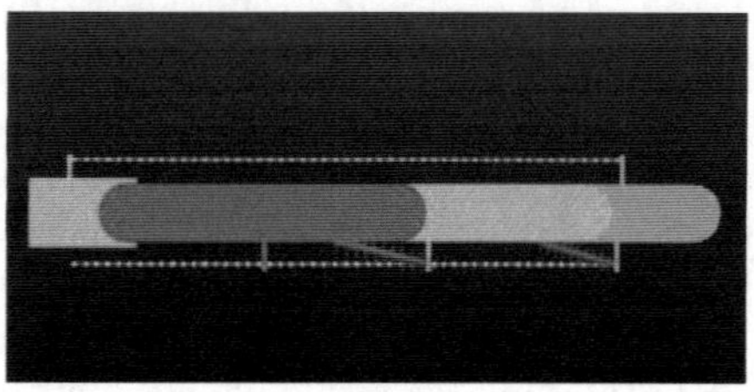

Figure 10. Index finger model in Adams.

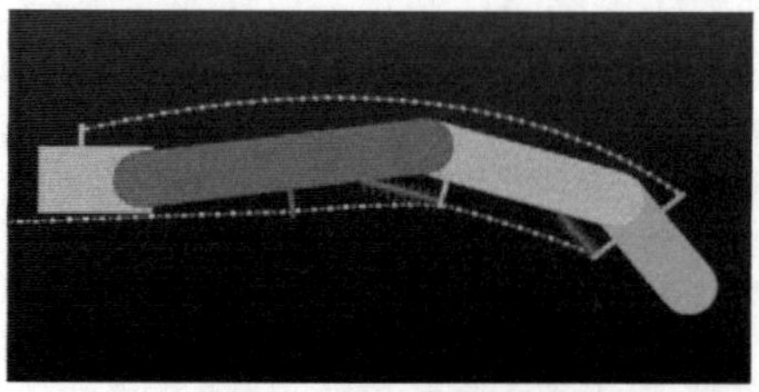

Figure 11. Dynamic simulation process of index finger.

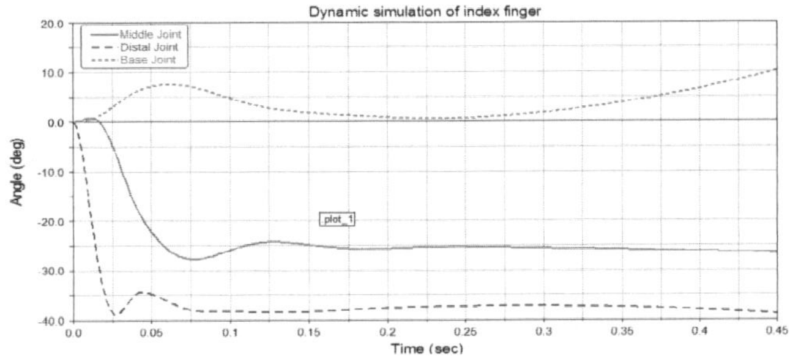

Figure 12. Simulation result.

When the driving force in the case of 4.5*N*, the angle of distal joint, middle joint and proximal joint is range of 0~38 degrees, 0~27 degrees and 0~8 degrees. The motion range of each joint accords with the actual movement condition of finger basically, and the running rule accords with the actual movement rule of human finger basically.

4. Conclusion

Aiming shortage at the existing structure of rehabilitation gloves, a new type of rehabilitation robot glove is designed. The flexible driving device and flexible structure are adopted to ensure the smooth movement of the rehabilitation process. At the same time, the simulation analysis is carried out using Adams, which lays a foundation for control research. The springs are adopted to imitate the muscles of human hands, which acquires a good result and realized simulation of the movement of the hand. In the future, the robot modeling with human muscles will be studied in order to get control model accurately.

5. References

[1] Wege A, Kondak K and Hommel G 2005 mechanical design and motion control of a hand exoskeleton for rehabilitation. Mechatronics and Automation. *IEEE International Conference (2005)* 1 p155-159.

[2] Wege A and Hommel G 2005 development and control of a hand exoskeleton for rehabilitation of hand injuries. *IEEE/RSJ international Conference on Intelligent Robots and Systems (2005)*, p3046-3051.

[3] Fu Y, Wang P and Wang S 2007 design and development of a portable exoskeleton based CPM machine for rehabilitation of hand injuries. *IEEE International Conference on Robotics and Biomimetics (2007)* p1476-1481.

[4] Liu H S 2007 Structure design and analysis of mechanical hand for rehabilitation. *Harbin Institute of Technology Doctoral Dissertation.*

[5] Fontana M, Dettori A and Salsedo F 2009 Mechanical design of a novel Hand Exoskeleton for accurate force displaying. *IEEE International Conference on Robotics and Automation (2009)* p1704-1709.

[6] Chiri A, Vitiello N and Giovacchini F 2012 Mechatronic design and characterization of the index finger module of a hand exoskeleton for post-stroke rehabilitation. *IEEE/ASME Transactions on Mechatronics (2012)* **17** p884-894.

[7] Patar, Mohd Nor Azmi Bin Ab, T. Komeda, and J. Mahmud 2014 Force assisted hand and finger device for rehabilitation. *IEEE International Symposium on Technology Management and Emerging Technologies (2014)* p133-138.

6. Acknowledgments

This work is supported by Shandong key research and development program (public welfare special) (Grant NO.2017GGX30103) and The Fundamental Research Funds of Shan dong University (Grant NO.26010275611010).

Automatic Modeling and Simulation of Modular Robots

C Jiang, H Wei and Y Zhang

School of Mechanical Engineering and Automation, Beihang University, Beijing 100191, China

Abstract. The ability of reconfiguration makes modular robots have the ability of adaptable, low-cost, self-healing and fault-tolerant. It can also be applied to a variety of mission situations. In this manuscript, a robot platform which relied on the module library was designed, based on the screw theory and module theory. Then, the configuration design method of the modular robot was proposed. And the different configurations of modular robot system have been built, including industrial mechanical arms, the mobile platform, six-legged robot and 3D exoskeleton manipulator. Finally, the simulation and verification of one system among them have been made, using the analyses of screw kinematics and polynomial planning. The results of experiments demonstrate the feasibility and superiority of this modular system.
Keywords: modular robots; reconfiguration; screw theory

1. Introduction

Traditional robots are developed based on specific tasks [1]. Therefore, every robot can only adapt to a certain range of workspace. The disadvantages of poor flexibility, poor self-repairing, low fault tolerance and high cost make them more and more unsuitable for the changes of market. Meanwhile, what people really need is a programmable multifunctional flexible device that can meet the demands of different tasks by changing its own program.

So people have to develop the reconfigurable modular robot system, including static reconfigurable modular robot and dynamic reconfigurable modular robot. Reconfigurable modular robot system is composed of a set of modules with different size and performance characteristics, using these modules can quickly build a variety of different configurations. Because of these advantages, many scientific research institutions, universities and companies are engaged in the aspects of researching, designing and manufacturing of modular robots. But there are some shortcomings in the existing modular robot platform, such as single module types, less construction of robot systems, etc.

In this paper, a modular robot system based on the screw theory and modular theory was designed. Although the type of joint modules contained in this system is simple, we can achieve completely different movements and build different types of robots by combining different auxiliary modules.

2. Modular Robot Platform

IOP Conf. Series: Materials Science and Engineering **320** (2018) 012003 doi:10.1088/1757-899X/320/1/012003

2.1. Modular Robot System

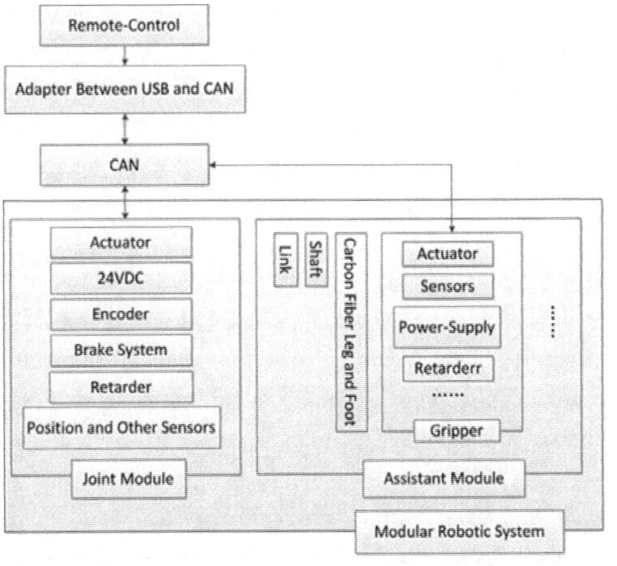

Figure 1. The Modular Robot System.

The modular system platform we designed, shown in figure 1, has many advantages [2]. First of all, similar modules consist of the same compositions, which means it is easy to assemble and disassemble, making it easier to upgrade and maintain. Secondly, it has a larger self-respect of loads than before. Thirdly, the method of hollow hole routing is adopted to eliminate the unsafe hidden troubles caused by the outer edges. And then, the robot configuration constructed with this module library use the least number of modules and module types. Next, the feature-rich API and compatibility between Windows and Linux systems will help users to redevelop. Finally, the modification of control algorithm in motion control is based on ROS, which is very simple.

2.2 the Partition of Modules

Currently, there are many ways to divide modules. Benhabib [3] divided the module system into four parts based on the angle of the mechanical, including joint module, connecting rod module, terminal execution module and basic unit module. Paredis [4] proposed the hardware module partition method. In this manuscript, the module library is divided into joint module and auxiliary module simply, and auxiliary module is a functional component to realize different robot system.

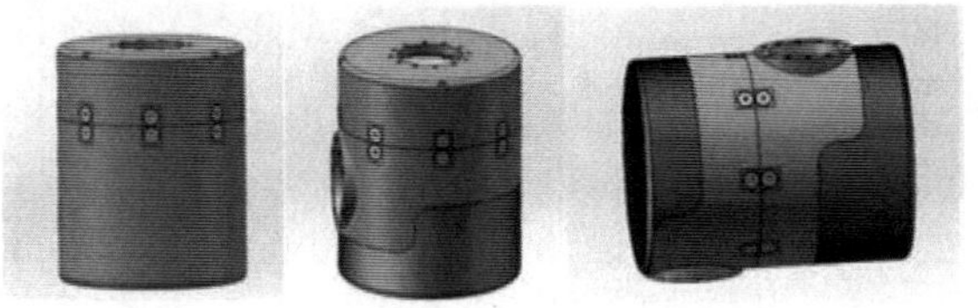

Figure 2. Joint Modules.

Figure 3. Auxiliary Modules.

A set of joint modules designed by this module library is shown in figure 2, and various auxiliary modules are shown in figure 3. All joints are rotatable joints. The upper and lower surfaces can rotate relative to 360 degrees continuously. When adding the appropriate front and rear covers on the joint, the non-parallel output can be achieved. Auxiliary modules are functional units designed to extend the functions of joint modules. Listed as figure 3, including wheel of mobile platform, bracket and assembly finger of manipulator, the connecting linkage of the manipulator, carbon fiber leg and motherboard of six-legged robot. The actual modular robot platform contains more modules.

In mechanical structure, the connection between joins and auxiliaries is realized by card ring and positioning pin, by which quick loading and unloading can be achieved. On the other hand, a CAN bus is used in the module communication system.

3. Configuration Design of Modular Robot

The configuration design is a problem to choose the optimal configuration from different modular combinations based on a specific task, which belongs to a kind of optimization analysis. It can be defined as designing complete robot configurations according to certain optimization criteria, by selecting suitable joint modules and auxiliary modules from the designed module library.

3.1 the Expression Method of Incidence Matrix Based on Graph Theory

The information of the number of models, the type of modules and the connection orientation in the configuration, can be expressed by applying graph theory and the Incidence matrix expression method. Because the joint modules in this module library are all rotating joints, the list of modular types in the incidence matrix is removed. The configuration, schematic diagram and correlation matrix of five degrees of freedom robot are shown in figure 4.

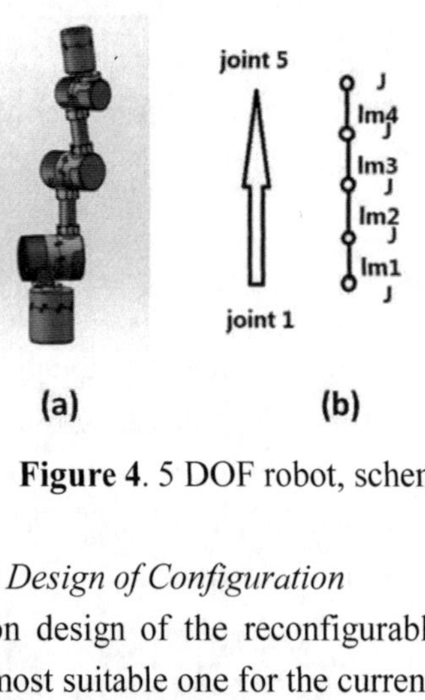

$$asm = \begin{bmatrix} +z & 0 & 0 & 0 & J \\ -z & +z & 0 & 0 & J \\ 0 & -z & +z & 0 & J \\ 0 & 0 & -z & +z & J \\ 0 & 0 & 0 & -z & J \\ lm_1 & lm_2 & lm_3 & lm_4 & \backslash \end{bmatrix}$$

(a) (b) (c)

Figure 4. 5 DOF robot, schematic diagram and correlation matrix.

3.2 Optimization Design of Configuration

The configuration design of the reconfigurable robot is guided by the target task. It refers to the selection of the most suitable one for the current task among various robot configurations assembled in the module library [5]. The target task is to realize the trajectory of the robot in Cartesian space. Therefore, the configuration design processing can also be defined as: according to the given position information of the robot in space, the optimal configuration satisfying the conditions is obtained through the constraints including configuration variables and target evaluation functions we set.

The process of configuration design of modular robot is shown in figure 5.

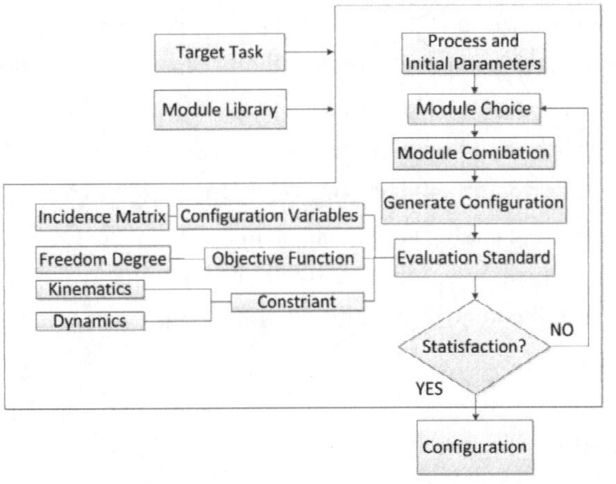

Figure 5. Configuration Design Processing.

For a given configuration requirement, the system will first select module configuration. Then it will traverse all configuration module libraries through the evaluation standard to judge. Finally it could get the optimal configuration in freedom, kinematics and dynamics, and output it to the user.

4. Modular Robot Configuration

According to the above configuration design method, the robot module library is used to build different kinds of robot system.

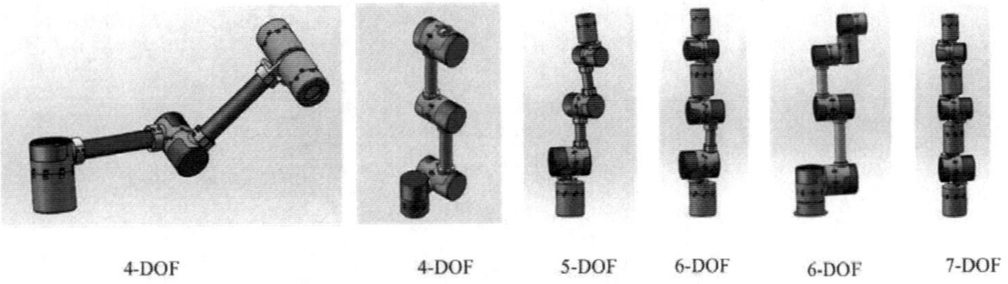

Figure 6. Industrial Manipulator Arms.

The mechanical arm built by the joint modules and link modules in the modular library is shown in figure 6. By using different types of joints to implement different output from input, we can control the pose of the end of the robotic arms. At the same time, we can add cameras, grippers and drills to the end of the manipulators to realize image-feedback, object-grabbing and industrial plastic-working. The robotic arm can be mounted on a mobile platform to enable remote wireless visual working, thus keeping the operator away from dangerous situations.

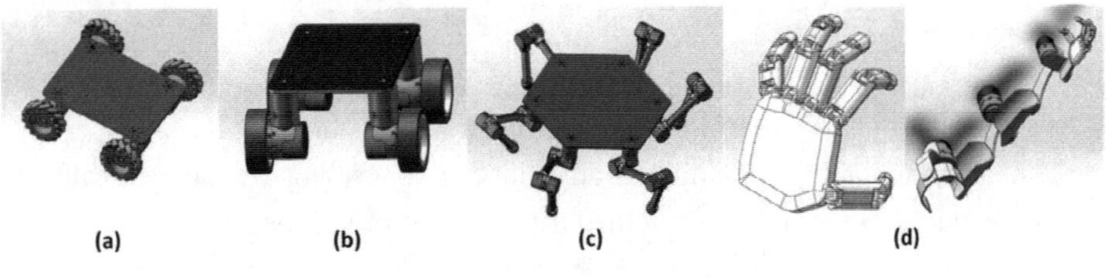

Figure 7. Some Robot Configurations.

Other configurations, such as wheeled mobile robot platform, six-legged crawling robot and wearable exoskeleton manipulator, are shown in figure 7. The mobile platforms, with integrating driving device, retarder and braking in its modules, can easily realize the forward- moving, deceleration and braking of the car. Thus, the car in figure b is more likely to pass roads which has height limit or go over obstacles. The six-legged crawling robot, as shown in figure c, uses carbon fiber linkages as the legs and hemispherical carbon fiber parts as foots. The controller is installed on the motherboard and the joints and legs are controlled through the CAN bus. We can achieve the goal of a six-legged robotic movement trajectory from parallel to vertical. Wearable exoskeleton manipulator, as shown in figure d, can do some work in place of human arm, such as automatic machine tools, automatic production lines, or medical treatment.

5. Modular Robot Simulation

Modular robots can be restructured according to task requirements, and each configuration of the robot corresponds to different mathematical models. At present, the most common method of kinematics is D-H modeling [6], but this method is very dependent on configuration. It means that the D-H model has to be rebuilt every time when the configuration changes, which is very complicated. Thus it is meaningful to present a more rapid general kinematics modeling method.

5.1 Inverse Kinematics

In this manuscript, the Screw analysis method of modular robot is put forward based on the combination of rigid body motion laws and mathematical methods, including of Li's group and Li's algebra, Screw theory and matrix operation. Therefore inverse kinematics analysis based on Screw is shown in figure 8.

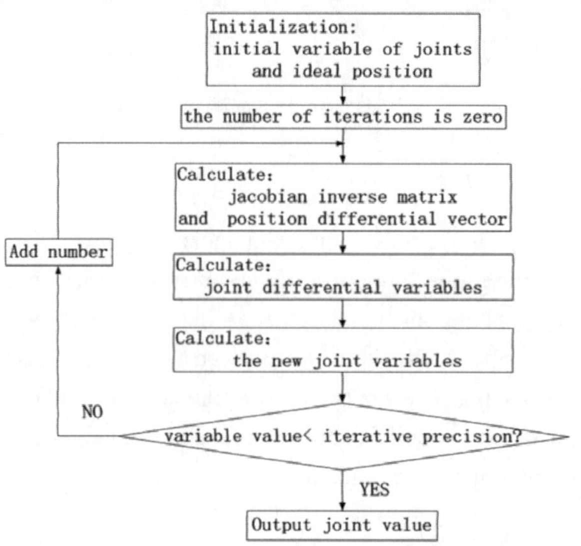

Figure 8. Inverse Kinematics Algorithm.

According to the formula of POE and the iterative inverse algorithm, we can write the MATLAB program to get the graphical user interface of the forward and inverse kinematics.

5.2. Simulation and Verification

In this manuscript, the trajectory planning method of polynomial interpolation in joint space is carried out [7]. By MATLAB to simulate the 5 DOF modular robots [8] and verify the method of the kinematic algorithm and the motion planning algorithm, the end position of the manipulator model was shown in figure 9.

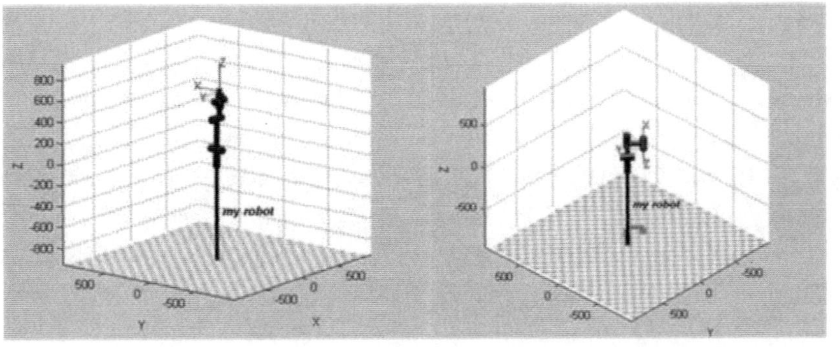

(a) Initial position (b) Expected position

Figure 9. the Initial Position and Expectation of the Manipulator.

In the simulation of the manipulator moving from the initial position to the terminal position, the pose trajectory of the end is shown in figure 10.

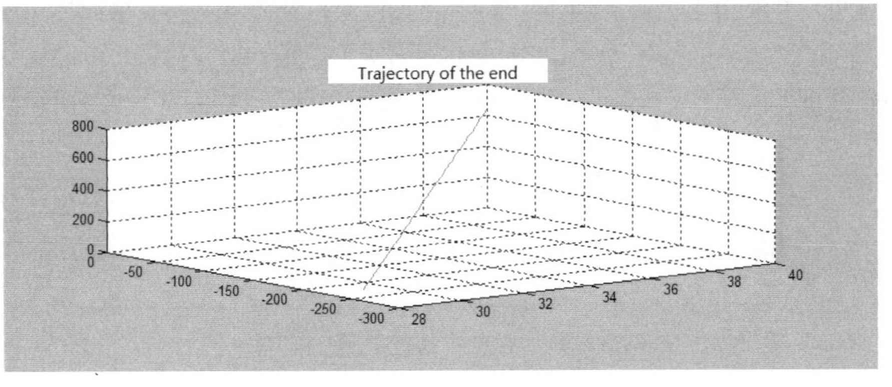

Figure 10. Pose Trajectory of the End.

6. Summary

In this manuscript, a kind of robotic platform based on module is designed. Firstly, the system architecture of modular library, the combination method of joint modules and auxiliary modules are presented. Then, with using this modular platform to develop several kinds of robots, including industrial robot arms, mobile platform robots, hexapod crawling robot and 3D exoskeleton manipulator, the configuration optimization method is given. In the end, with the kinematics analysis is carried out using screw theory and the trajectory planning is carried out using polynomial theory. The robot system is simulated and verified. The results of experiment show that the modular system is universal.

Further studies, such as double-programming environment, will be worked on. We hope a robot based on the module library will be compatible with both Linux and Windows, so that users can get the movement they want according to their own habits. We are also looking into the field of teaching robots, to make robots meet user's demands by imitating their actions.

7. Reference

[1] Breazeal C 2017 Social robots: from research to commercialization *ACM/IEEE International Conference on Human-Robot Interaction* p 1

[2] Andani M E, Bahrami F, Maralani P J and Ijspeert A J 2009 Modem: a multi-agent hierarchical structure to model the human motor control system *Biological Cybernetics* vol 101 pp 361-377

[3] Benhabib B, Zak G and Lipton M G 2010 A generalized kinematic modeling method for modular robots *Journal of Robotic Systems* vol 6 no 5 pp 545-571

[4] Paredis C J J, Brown H B and Khosla P K 1996 A rapidly deployable manipulator system *IEEE Proceedings of International Conference on Robotics and Automation* Vol 21 pp 1434-1439

[5] Dennl C, Ziener D and Teich J 2012 On-the-fly composition of fpga-based sql query accelerators using a partially reconfigurable module library *IEEE 20th International Symposium on Field-Programmable Custom Computing Machines*

[6] Sun J D, Cao G Z, Li W B, Liang Y X and Huang S D 2017 IEEE Analytical inverse kinematic solution using the D-H method for a 6-DOF robot *International Conference on Ubiquitous Robots and Ambient Intelligence* pp 714-716

[7] Luo L P, Yuan C, Yan R J, Yuan Q, Wu J and Shin K S 2015 Trajectory planning for energy minimization of industry robotic manipulators using the lagrange interpolation method *International Journal of Precision Engineering & Manufacturing* vol 16 no 5 pp 911-917

[8] Vennishmuthu V 2014 Stability simulation of six dof manipulators using adams and matlab *IOSR Journal of Engineering* vol 4 no 3 pp 09–12

IOP Conf. Series: Materials Science and Engineering **320** (2018) 012004 doi:10.1088/1757-899X/320/1/012004

Kinematics Control and Analysis of Industrial Robot

Tongbo Zhu[1], Fan Cai[1], Yongmei Li[1] and Wei Liu[1]

[1] Minnan university of science and technology, Provincial Key Laboratory of industrial robot measurement and control,Rapid die manufacturing, Quanzhou Fujian 362700,China

Abstract. The robot's development present situation, basic principle and control system are introduced briefly. Research is mainly focused on the study of the robot's kinematics and motion control. The structural analysis of a planar articulated robot (SCARA) robot is presented,the coordinate system is established to obtain the position and orientation matrix of the end effector,a method of robot kinematics analysis based on homogeneous transformation method is proposed, and the kinematics solution of the robot is obtained.Establishment of industrial robot's kinematics equation and formula for positive kinematics by example. Finally,the kinematic analysis of this robot was verified by examples.It provides a basis for structural design and motion control.It has active significance to promote the motion control of industrial robot.

1. Introduction

The robot is the most typical electromechanical integrated digital equipment with high added value and wide range of applications. As an emerging industry of information society and supporting technology, advanced manufacturing will play a more and more important role in the future production and social development [1].In recent years, the robot industry has maintained a rapid growth trend. Robot's manufacturing demand market potential, and usher in an excellent opportunity for development of Chinese robot automation industry. Chinese industrial robot market developed rapidly, accounting for about 1/3 of the global market share, is the world's largest industrial robot application market.

It pays a great attention to the research of robot technology in the national high technology research and development plan, the National Natural Science Foundation and the national science and technology special program in china. It has a great expectations for the future development of robot technology in the industry at home and abroad. Thus, the technology of robot is one of the foundations for the development of high-tech and emerging industries in the future. It is of great significance to the national economy and national defense construction[2].

2. The basic working principle of industrial robot

Industrial robot is a kind of production equipment, whose basic function is to provide the movement and power required by homework. The basic working principle is that the movement function and technical requirements of the hand operation can be realized automatically through the movement of each moving component on the operating machine. Therefore, in the basic function and working principle, the industrial robot and the machine tool have the same place:the end effector of the two has the position change requirement, and the position change requirement of the end effector is realized by coordinate motion. Of course, the robot also has its unique requirements, mainly in the form of joint movement. At the same time, the flexibility of the robot is required to be high, and its stiffness and precision is required to be low.

3. Motion control system of industrial robot

Motion control system is one of the key components to determine the performance of industrial robots. The core technology of improving the characteristics of motion control system and enterprise competitiveness. Therefore, it is of great significance to study a motion control system which is stable, efficient and suitable for industrial robot operation. [3]

The motion control system of industrial robot , which includes program interpreter, motion interpolator, servo driver and some other auxiliary modules, is composed. The motion interpolator is equivalent to the part of the human brain intelligence, motion control command which receives the user program instruction interpreter output, and according to the requirements of the directive with technology, careful planning position or trajectory parameters, to drive the server. For limited space, it studys kinematics and motion control of industrial robot in the following Words.[4]

4. kinematics and motion control of industrial robot

4.1 The kinematics

It studys the time history of the position, velocity and acceleration for the kinematics of industrial robot. At present, the industrial robot which has the six degree of freedom(SDOF) is widely used in the equipment manufacturing industry, and the modeling of the kinematics of the six degree of freedom robot has been more mature.But the robot's running, trajectory planning, and the dynamics of the robot is determined by the robot kinematics modeling[5]. Therefore, it is very important to establish kinematic model which is a simple solution method for robot system. Therefore, it is very important to establish and solve a simple kinematic model for robot system. The solution method is simple kinematic model.[6,7]

4.1.1 Robot's kinematics equation. The coordinate system is established according to the robot connecting rod, and the relative relation between the coordinate systems is described by the homogeneous transformation equation. This homogeneous transformation matrix is usually called the A matrix.[8]

If the matrix A_1 is used to represent the homogeneous transformation of the first link coordinate system with respect to the fixed coordinate system, the pose is T_1

$$T_1 = A_1 \cdot T_0 \tag{1}$$

Formula: T_0 is the homogeneous matrix expression of the fixed coordinate system,

$$T_0 = \begin{bmatrix} 1 & 0 & 0 & 0 \\ 0 & 1 & 0 & 0 \\ 0 & 0 & 1 & 0 \\ 0 & 0 & 0 & 1 \end{bmatrix} \tag{2}$$

If the matrix A_2 represents the homogeneous transformation of the second link coordinate systems with respect to the first link coordinate system, then the pose T_2 can be represented by the product of A_1 and A_2, and the A_2 should be multiplied by right

$$T_2 = A_1 \cdot A_2 \tag{3}$$

Similarly, if the matrix A_3 represents the homogeneous transformation of the third link coordinate systems with respect to the second link coordinate systems, there is:

$$T_3 = A_1 \cdot A_2 \cdot A_3 \tag{4}$$

In this way, the following T_6 matrices:

$$T_6 = A_1 \cdot A_2 \cdot A_3 \cdot A_4 \cdot A_5 \cdot A_6 \tag{5}$$

This formula represents a multiplicative transform matrix between the right hand coordinate system from the fixed reference frame of the connecting rod coordinate system, T_6 said the hand coordinate system with respect to a fixed reference position, we call type (5) for the kinematics equation of the robot. Formula (5) the result of calculation T_6 is a following (4 * 4) matrix:

$$T_6 = \begin{Bmatrix} n_x & o_x & a_x & p_x \\ n_y & o_y & a_y & p_y \\ n_z & o_z & a_z & p_z \\ 0 & 0 & 0 & 1 \end{Bmatrix} \tag{6}$$

In the formula, the first three columns represent the gesture of the hand, and the fourth column indicates the position of the hand.

4.1.2 positive kinematics. The positive kinematics mainly solves the problem of the establishment of robot kinematics equation and the solution of the hand position and pose.[9] The following are the methods of establishing the robot kinematics equation and two examples.

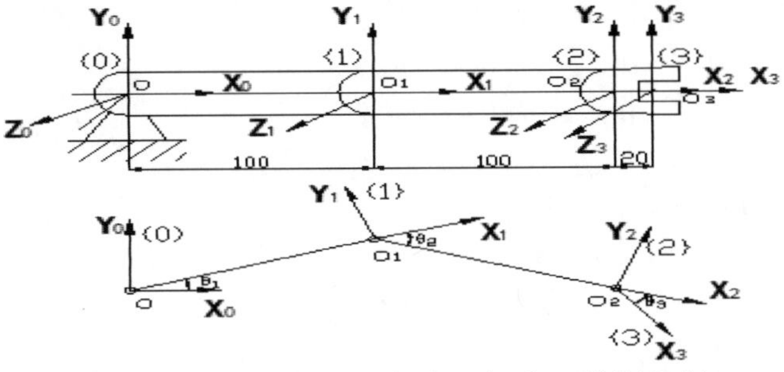

Figure 1. Planar articulated robot (SCARA)

As shown, a planar joint robot (SCARA), {0},{1},{2} and {3}respectively represents a shoulder joint, an elbow joint, and a wrist joint[10]. The center of the circle is O, O_1, $O_2$and O_3. The corresponding coordinate system is XYZ, $X_1Y_1Z_1$, $X_2Y_2Z_2$ and $X_3Y_3Z_3$.The mechanical structure of such robots is that the axes of the three joints are parallel to each other.The coordinate system (o) and the connecting rod 1, the connecting rod 2 and the connecting rod 3 are respectively located at the joint 1, the joint 2, the joint 3 and the hand center.The D in the linkage parameter is the variable, and the other parameter J is the root. All constants. Considering that the axis of the joint is parallel and that the connecting rods are all in one plane, the parameters of the planar joint robot (SCARA) robot link are shown in Table 1.

Table 1. Connecting rod parameters of planar articulated robot (SCARA).

connecting rod	angle (variables) θ	Distance between two connecting rods d	connecting rod lengtha	Torsion angle of connecting rod α
connecting rod 1	θ_1	$d_1=0$	$a_1=l_1=100$	$\alpha_1=0$
connecting rod 2	θ2	$d_2=0$	$a_2=l_2=100$	α2=0
connecting rod 3	θ3	$d_3=0$	$a_3=l_3=20$	α3=0

The kinematics equations of the planar joint robot are:

$$T_3 = A_1 \cdot A_2 \cdot A_3 \tag{7}$$

Formula A1 represents the homogeneous transformation matrix of the connecting rod 1 coordinate system {1} with respect to the fixed coordinate system {0};A2 represents the homogeneous transformation matrix of the connecting rod 2 coordinate system {2} with respect to the fixed coordinate system {1};A3 represents the homogeneous transformation matrix of the connecting rod 3 coordinate system {3} with respect to the fixed coordinate system {2}; Hence:

$$A_1 = Rot(z_0, \theta_1) Trans(l_1, 0, 0) \tag{8}$$

$$A_2 = Rot(z_1, \theta_2) Trans(l_2, 0, 0) \tag{9}$$

$$A_3 = Rot(z_2, \theta_3) Trans(l_3, 0, 0) \tag{10}$$

$$A_3 = \begin{bmatrix} c\theta_3 & -s\theta_3 & 0 & 0 \\ s\theta_3 & c\theta_3 & 0 & 0 \\ 0 & 0 & 1 & 0 \\ 0 & 0 & 0 & 1 \end{bmatrix} \begin{bmatrix} 1 & 0 & 0 & l_3 \\ 0 & 1 & 0 & 0 \\ 0 & 0 & 1 & 0 \\ 0 & 0 & 0 & 1 \end{bmatrix} = \begin{bmatrix} c\theta_3 & -s\theta_3 & 0 & l_3 c\theta_3 \\ s\theta_3 & c\theta_3 & 0 & l_3 s\theta_3 \\ 0 & 0 & 1 & 0 \\ 0 & 0 & 0 & 1 \end{bmatrix} \tag{11}$$

$$A_3 = \begin{bmatrix} c\theta_3 & -s\theta_3 & 0 & 0 \\ s\theta_3 & c\theta_3 & 0 & 0 \\ 0 & 0 & 1 & 0 \\ 0 & 0 & 0 & 1 \end{bmatrix} \begin{bmatrix} 1 & 0 & 0 & l_3 \\ 0 & 1 & 0 & 0 \\ 0 & 0 & 1 & 0 \\ 0 & 0 & 0 & 1 \end{bmatrix} = \begin{bmatrix} c\theta_3 & -s\theta_3 & 0 & l_3 c\theta_3 \\ s\theta_3 & c\theta_3 & 0 & l_3 s\theta_3 \\ 0 & 0 & 1 & 0 \\ 0 & 0 & 0 & 1 \end{bmatrix} \tag{12}$$

$$A_3 = \begin{bmatrix} c\theta_3 & -s\theta_3 & 0 & 0 \\ s\theta_3 & c\theta_3 & 0 & 0 \\ 0 & 0 & 1 & 0 \\ 0 & 0 & 0 & 1 \end{bmatrix} \begin{bmatrix} 1 & 0 & 0 & l_3 \\ 0 & 1 & 0 & 0 \\ 0 & 0 & 1 & 0 \\ 0 & 0 & 0 & 1 \end{bmatrix} = \begin{bmatrix} c\theta_3 & -s\theta_3 & 0 & l_3 c\theta_3 \\ s\theta_3 & c\theta_3 & 0 & l_3 s\theta_3 \\ 0 & 0 & 1 & 0 \\ 0 & 0 & 0 & 1 \end{bmatrix} \tag{13}$$

Therefore, it can be written out

$$T_3 = \begin{bmatrix} c_{123} & -s_{123} & 0 & l_3 c_{123} + l_2 c_{12} + l_1 c_1 \\ s_{123} & c_{123} & 0 & l_3 s_{123} + l_2 s_{12} + l_1 s_1 \\ 0 & 0 & 1 & 0 \\ 0 & 0 & 0 & 1 \end{bmatrix} \tag{14}$$

Formula:

$c_{123} = \cos(\theta_1 + \theta_2 + \theta_3); s_{123} = \sin(\theta_1 + \theta_2 + \theta_3); c_{12} = \cos(\theta_1 + \theta_2); s_{12} = \sin(\theta_1 + \theta_2); c_1 = \cos\theta_1;$
$s_1 = \sin\theta_1$

T3 is the result of multiply by A_1、A_2、A_3, represents the position and attitude of the hand coordinate system {3}.

$$T_3 = \begin{bmatrix} n_x & o_x & a_x & p_x \\ n_y & o_y & a_y & p_y \\ n_z & o_z & a_z & p_z \\ 0 & 0 & 0 & 1 \end{bmatrix} \tag{15}$$

Thus, the hand position (4 * 1) of the hand can be positioned as an array

$$P = \begin{bmatrix} p_x \\ p_y \\ p_z \\ 1 \end{bmatrix} = \begin{bmatrix} l_3 c_{123} + l_2 c_{12} + l_1 c_1 \\ l_3 s_{123} + l_2 s_{12} + l_1 s_1 \\ 0 \\ 1 \end{bmatrix} \tag{16}$$

The direction vectors n, o and a that represent the gesture of the hand are

$$n = \begin{bmatrix} n_x \\ n_y \\ n_z \\ 0 \end{bmatrix} = \begin{bmatrix} c_{123} \\ s_{123} \\ 0 \\ 0 \end{bmatrix} \tag{17}$$

$$o = \begin{bmatrix} o_x \\ o_y \\ o_z \\ 0 \end{bmatrix} = \begin{bmatrix} -s_{123} \\ c_{123} \\ 0 \\ 0 \end{bmatrix} \quad (18)$$

$$a = \begin{bmatrix} a_x \\ a_y \\ a_z \\ 0 \end{bmatrix} = \begin{bmatrix} 0 \\ 0 \\ 1 \\ 0 \end{bmatrix} \quad (19)$$

When the corner variable $\theta_1, \theta_2, \theta_3$ is timed, the specific value can be calculated[11]. As shown in the figure above. hypothesis, $\theta_1 = 30°, \theta_2 = -60°, \theta_3 = -30°$. Then the kinematics positive solution can be obtained according to the kinematics equation of the planar joint robot, that is, the position matrix of the hand is expressed as:

$$T_3 = \begin{bmatrix} 0.5 & 0.866 & 0 & 183.2 \\ -0.866 & 0.5 & 0 & -17.32 \\ 0 & 0 & 1 & 0 \\ 0 & 0 & 0 & 1 \end{bmatrix} \quad (20)$$

5. Summarize

Industrialization and industrial application of industrial robots is an important symbol to measure a country's scientific and technological innovation and high-end manufacturing development level.

To develop the robot industry is of great significance for creating new advantages in China's manufacturing, promoting industrial transformation and upgrading, speeding up the construction of manufacturing power and improving the living standards of the people.Establishment of industrial robot's kinematics equation and formula for positive kinematics by example. It has active significance to promote the motion control of industrial robot.

6. Acknowledgment

This work is sponsored by science and technology project of the Education Department of Fujian (JB13233) and the key laboratory of industrial robot measurement, control and die manufacturing (Fujian Education Section [2017] No. 8).

7. References

[1] Tianran Wang. Development of robot technology, [J]. robotics, 2017,38 (04): 385-386.

[2] Tongbo Zhu, Guangyu Zhu. Industrial robot, mechanical system kinematics and motion control, [J]. mechanical and electrical product development and innovation, 2016,29 (05): 15-17+58.

[3] Min Tan, Shuo Wang. Progress in robotics research [J]. Journal of automation, 2013,39 (07): 963-972.

[4] Yunfeng Gao,Mingrui Lv,Lun Zhou,Ruifeng Li.Analysis of kinematics of a five-axis hybrid manipulator [J]. Journal of Harbin Institute of Technology, 2014,46(7): 1-7.

[5] Peiwen Han, Jing Zhou, Lin Jiang, Jiyoung Zhang. Algorithm for welding [J]. manufacturing automation robot motion control, 2017,39 (07): 152-156.

[6] Dongdong Zhou,Guodong Wang,Juliang Xiao,Ying Hong. Design and kinematics analysis of new modular reconfigurable robot [J]. Chinese Journal of Engineering Design, 2016,23(01): 74-81.

[7] Yi Cao,Youlei Qin,Hai Chen,Hui Zhou,Baokun Li.Structural Synthesis of 3T2R Five-Degree-of -Freedom Hybrid Mechanism[J].Journal of Donghua University(English Edition), 2016, 35(3): 114 - 121

[8] Tianmiao Wang,Yong Tao.Technology status and industrial development strategy of industrial robots in China [J]. Chinese Journal of mechanical engineering, 2014,50 (09): 1-13.

[9] Yinhui Li,Yanwen Li. Dynamics analysis of milling robot based on SolidWorks and MATLAB.[J].Journal of Machine Design, 2016,33(12): 27-30

[10] Summer chain,Xiaohui Yu , Jiang Han, Xiaoqing Tian. Industrial robot UMAC control system design based on [J]. Journal of HeFei University of Technology , 2015,38 (08): 1009-1012.

[11] Yafeng Li, Jianghai Zhao, Xiaojian Zhang, Feng Xue .Research of the mechanism configuration and motion control for the cable-driven hybrid joint robot[J]. Manufacturing Automation 2016, 38 (08): 36-40+49.

IOP Conf. Series: Materials Science and Engineering **320** (2018) 012005 doi:10.1088/1757-899X/320/1/012005

Kinematics Modelling of Tendon-Driven Continuum Manipulator with Crossed Notches

Z X Yang, W L Yang and Z J Du

State Key Laboratory of Robotics and System, Harbin Institute of Technology (HIT), Harbin, China
Corresponding author: duzj01@hit.edu.cn

Abstract. Single port surgical robot (SPSR) is a giant leap in the development of minimally invasive surgical robot. An innovative manipulator with high control accuracy and good kinematic dexterity can reduce wound, expedite recovery, and improve the success rate. This paper presents a tendon-driven continuum manipulator with crossed notches. This manipulator has two degrees of freedom (DOF), which possesses good flexibility and high capacity. Then based on cantilever beam theory, a mechanics model is proposed, which connects external force and deformation of a single flexible ring (SFR). By calculating the deformation of each SFR, the manipulator is considered as a series robot whose joint numbers is equal to SFR numbers, and the kinematics model is established through Denavit-Hartenberg (D-H) procedure. In this paper, the total manipulator is described as a curve tube whose curvature is increased from tip to base. Experiments were conducted and the comparison between theoretical and actual results proved the rationality of the models.

1. Introduction

Minimally invasive surgery (MIS) is featured by small trauma, less bleeding, light pain, and quick recovery [1]. As the second generation surgical robot, SPSR needs only one wound on patient's body instead of 3 to 5 wounds necessarily in traditional MIS. The limitation of wound reduces the injuries to people, but innovative manipulators need to be designed to perform essential motion [2].

Mainstream SPSRs can be divided into two types, one contains traditional discrete joint arm, and the other one adopts novel continuum manipulator. The first type has high positional accuracy and easy to control, but it needs relatively large wound. Wortman [3] devised an in vivo robot which contains two operation arms, whose motors are on the joints. Petroni [4] developed the SPRINT surgical platform including image system. Quaglia [5] presented SPRINT 2.0 with motors installed outside and tendons adopted to transmit force. The second type has miniature size and flexible movement, but control is complex and laden capacity is limited. Webster [6] and Wu [7] proposed the snake robots made of concentric pre-curved tubes. Xu [8] and Li [9] introduced a kind of manipulator in which the backbone is a continuous elastic structure in the centre and drivers are tendons distributed uniformly around. Du [10] and Gao [11] built continuum manipulators with different notches.

Continuum manipulators have excellent flexibility and biocompatibility, but the existing ones still have the problem of poor control accuracy and insufficient laden capability. This paper designs a continuum manipulator with high rigidity, and a relatively precise kinematics model is established for control.

2. Overall design of the continuum manipulator

Analysing of existing SPSRs, this section presents a novel structure, a notched flexible continuum manipulator driven by tendons.

2.1. Design of the structure

Flexible hinges have many advantages such as no back, friction, gap and wear, small size and so on, which is widely used in many fields [12]. These features are also expected for the manipulator of SPSR, so this paper designs SFR based on the idea of flexible hinge.

The crossed notches are fabricated on hyperelastic nitinol tube through wire cut electrical discharge machining (WEDM), and the continuum manipulator is a combination of several crossed SFRs whose bending planes are perpendicular to each other. The continuum manipulator is driven by two sets of drive tendons, each consisting of two tendons, which uniformly distribute in the tube. Each set of tendons controls forward and reverse bending on a plane. Combination of two sets can realize universal bending. The tendons are fixed at the top of the continuum manipulator and the flexible skeleton is installed internally to fit tendons, as shown in Figure 1.

2.2. Features of the continuum manipulator

The notched continuum manipulator achieves overall processing, and no connection friction extends the service life span. The manipulator has good flexibility and biocompatibility, which can avoid unnecessary damage to organs and issues during surgery. The tendons minimize the size, weight, and volume of the manipulator. Compared with our previous design, the crossed SFRs make one manipulator achieve two DOF, which improves control and movement dexterity. Also, according to the first-order natural frequency, the proposed manipulator overcomes insufficient rigidity which causes low capability. Figure 2 shows the results of finite element analysis by ANSYS, and each structure has four deformation units and similar key parameters.

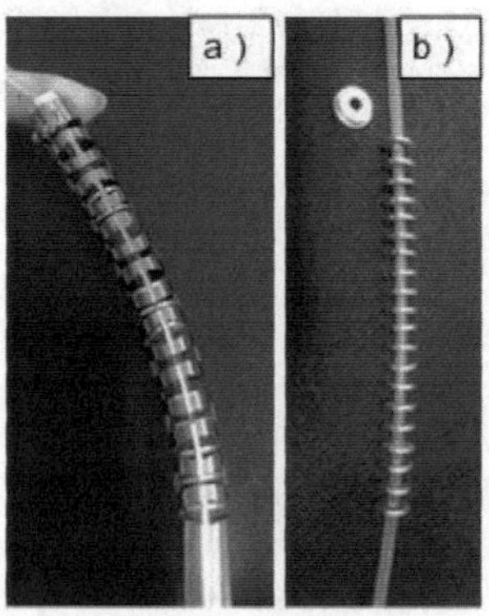

Figure 1. Prototype.

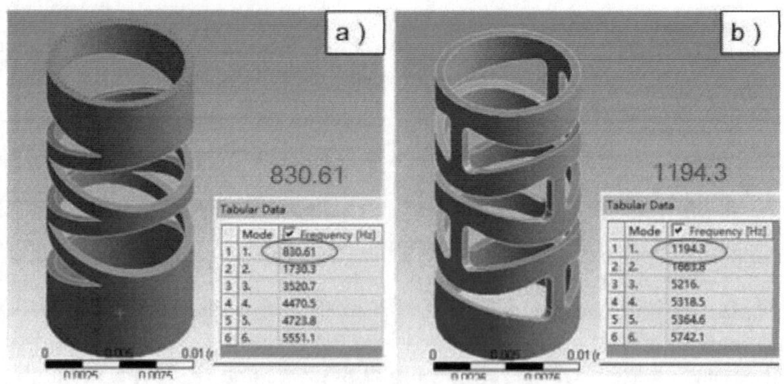

Figure 2. Results of stiffness analysis.

3. Mechanics modelling of continuum manipulator

The mechanical properties of continuum manipulator are deduced. A SFR is divided into rigid region and elastic region. Stiffness characteristics are built using the theory of cantilever beam, and deformation for SFR is calculated by the iterative algorithm.

3.1. Analysis of SFR stiffness characteristics

Because of the complex shape of SFR, the following assumptions are put forward: deformation mainly concentrates on the rounded straight flexible part, and deformation outside this part is ignored; interference between different SFRs can be neglected because deformation is very small; the SFR is assumed to be fixed at one end and subjected to torque and force on the other end. The flat projection and structural parameters of SFR are shown in Figure 3. R is the outer radius of tube, b is the wall

thickness, t is the minimal width of flexible hinge, l is the length of straight hinge, r is the radius of rounded circle and h is the vertical distance between two SFRs.
According to the actual loading situation, the main external forces on SFR are exerted by tendons. To describe stiffness characters, two variables are introduced: bending stiffness K_θ is defined as the ratio of bending moment to corresponding bending angle; tension/compression stiffness K_l is defined as the ratio of the tension/compression force to corresponding translation displacement [13]. E is the elastic modulus, which of nitinol is affected by the processing method, so actual elastic modulus is obtained by parameter identification. According to statics, K_θ and K_l of FNR can be calculated as follows.

$$K_\theta = \int_{-l/2}^{l/2} \frac{Ebt^3}{6} dx + \int_{-\pi/2}^{\pi/2} \frac{Eb(t+2r-2r\cos\alpha)^3}{6r\cos\alpha} d\alpha \tag{1}$$

$$K_l = \frac{2Ebt}{l} + \int_{-\pi/2}^{\pi/2} \frac{2Eb(2r+\mathrm{t}-2r\cos\alpha)}{r\cos\alpha} d\alpha \tag{2}$$

Using calculus theory, the comparison between bending deformation and translation displacement under the same force is calculated, and the results show that the former one is much bigger than the latter one. To simplify the model, the following calculation assumes that the deformation only occurs on the bending plane.

3.2. Angle deformation of SFR
Angle deformation for the ith notch θ_i is calculated using the iterative algorithm, and the computing process is as follows: for initial calculation, assume that there is no lateral pressure, then the bending angle $\theta_i(0)$ for SFR is obtained; at this time lateral pressure $F_{Ni}(0)$ and friction force $F_{fi}(0)$ can be calculated; because of the aforementioned two forces, new deformation is derived, which is written as $\theta_i(n)$ for the nth calculation; the difference between two adjacent calculations is viewed as objective functions; repeat the above process until the objective function tends to zero so the final angle deformation θ_i is calculated. Here μ is the friction coefficient which is given by experience. The force diagram of SFR is shown in Figure 4.

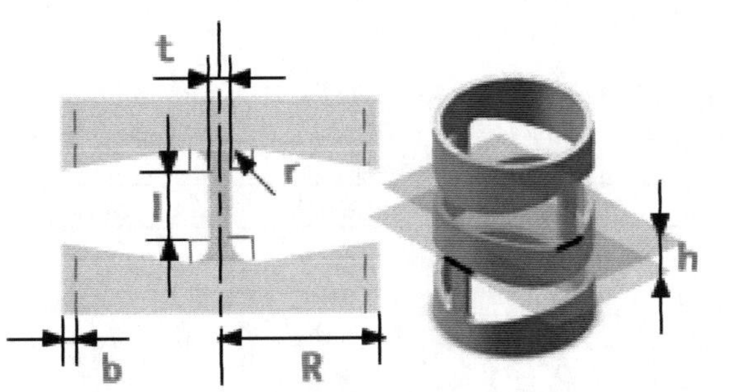

Figure 3. Flat projection of SFR.

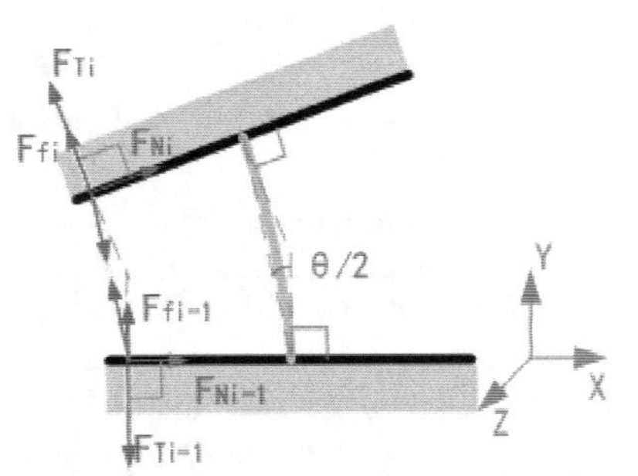

Figure 4. Force diagram of SFR.

Equivalent bending moment equals to:

$$M(n) = F_x L_y + F_y L_x \tag{3}$$

Where F_X and F_Y are the decomposition of all the forces in Cartesian coordinate system, L_X and L_Y are the corresponding force arm, all of these related to deformation. The bending angle for the nth iterative algorithm can be calculated according to $M(n)$ and K_θ. Due to frictional forth and lateral pressure, tendon force varies in different segments. Angle deformation for all SFRs can be obtained according to the varied equivalent tendon forces.

4. Kinematics modelling of continuum manipulator

The kinematics model of continuum manipulator proposed in this paper regards the whole manipulator as a series robot, which divides the manipulator into several equivalent rigid links and joints. The initial subcoordinate system is established at the end of base SFR, the final subcoordinate system is established at the top of tip SFR, and other subcoordinate systems are established at the intersection of two tangents of adjacent SFRs. The coordinate system is shown in Figure 5.

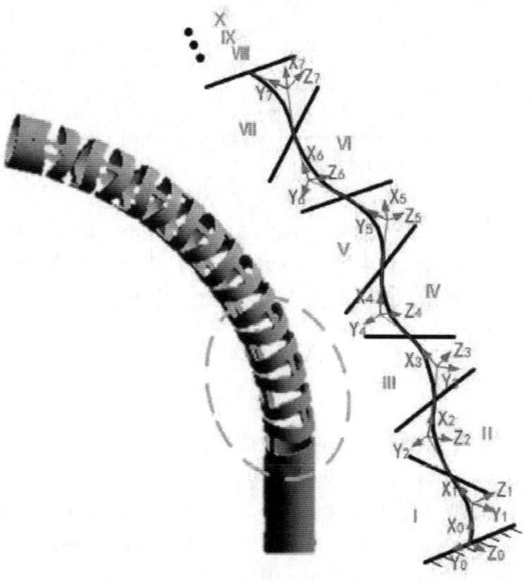

Figure 5. Coordinate system of the continuum manipulator.

The deformation angle of each SFR can be obtained according to the mechanics model in section 3, and` the D-H method can be applied to build kinematics model. This paper uses the Craig rule and D-H parameters are shown in Table 1. α_i is the twist angle of the link, l_i is the length of the link, d_i is the offset of the link, and β_i is the rotation angle of the link.

Table 1. D-H parameter table.

Parameters	Number of notches					
	1	2	3	…	2n-1	2n
α_i	π/2	-π/2	π/2	…	π/2	-π/2
l_i	l_1	l_2	l_3	…	l_{2n-1}	l_{2n}
d_i	0	0	0	…	0	0
β_i	θ_1	θ_2	θ_3	…	θ_{2n-1}	θ_{2n}

Where β_i is equal to the angle deformation θ_i of the ith SFR, and l_i can be calculated as follows:

$$l_i = \begin{cases} h + \beta_i^{-1}(l+2r)\tan(\beta_i/2) & i=1 \\ h + \beta_{i-1}^{-1}(l+2r)\tan(\beta_{i-1}/2) + \beta_i^{-1}(l+2r)\tan(\beta_i/2) & 1 < i \le 2n \\ \beta_{i-1}^{-1}(l+2r)\tan(\beta_{i-1}/2) & i = 2n+1 \end{cases} \tag{4}$$

The transformation matrix between adjacent equivalent rigid joints can be expressed as:

$$ {}^{i-1}\mathrm{T}_i = \begin{bmatrix} \cos\beta_i & -\sin\beta_i & 0 & l_{i-1} \\ \sin\beta_i\cos\alpha_{i-1} & \cos\beta_i\cos\alpha_{i-1} & -\sin\alpha_{i-1} & -d_i\sin\alpha_{i-1} \\ \sin\beta_i\sin\alpha_{i-1} & \cos\beta_i\sin\alpha_{i-1} & \cos\alpha_{i-1} & d_i\cos\alpha_{i-1} \\ 0 & 0 & 0 & 1 \end{bmatrix} \tag{5} $$

The overall transformation matrix of the continuum manipulator from base coordinate system $\{0\}$ to tip coordinate system $\{2n\}$ can be written as:

$$ {}^{0}T_{2n} = {}^{0}\mathrm{T}_1\,{}^{1}\mathrm{T}_2\,{}^{2}\mathrm{T}_3\cdots{}^{i-1}\mathrm{T}_i \tag{6} $$

The end position of the continuum manipulator can be obtained by equation (6). Calculated working space is shown in Figure 6.

5. Experiments and results

The test bed was constructed as shown in Fig.7. A prototype was machined for experiment, which had 8 notches in one bending direction (16 notches in total), and the effective length is 87.6mm. The multiple strand tendon adopted in experiment has a diameter of 0.45mm. The loading was exerted by the counterweight. Two CCD cameras were used in the experiment to take photos, and then angle deformation and spatial position were obtained by image processing using MATLAB software.

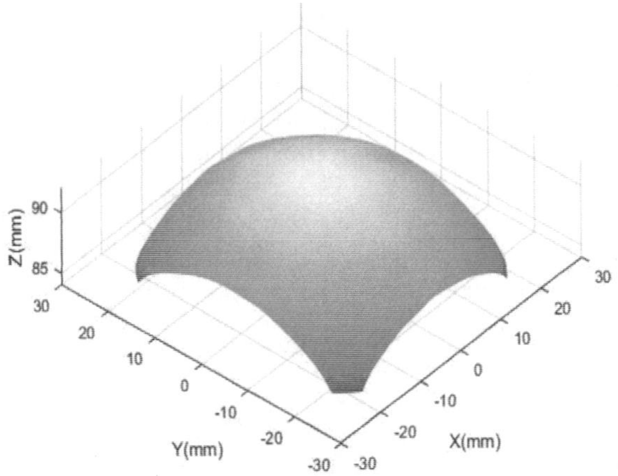

Figure 6. Working space.

Figure 7. Establishment of test bed.

5.1. Validation of mechanics model

One tendon was loaded each time, so the bending occurred in the plane. The load was applied from 0 to 2 kg by an increase of 0.2 kg, and then from 2 kg to 0 kg by a decrease of 0.2 kg. The actual bending angles under different loads were the average value of 5 cycles. The result between theoretical value and actual value is shown in Figure 8. The maximum deviation is 3.41%, which comes from model error, manufacturing error and image processing error.

5.2. Validation of kinematics model

Two tendons of different sets were loaded each time, so the bending occurs in the space. Different combinations of two tendons were applied. Each position two photos from different cameras were needed and the actual deformation displacement was obtained. The result between theoretical value and actual value is shown in Figure 9. The maximum deviation is 3.72%, which also comes from model error, manufacturing error and image processing error.

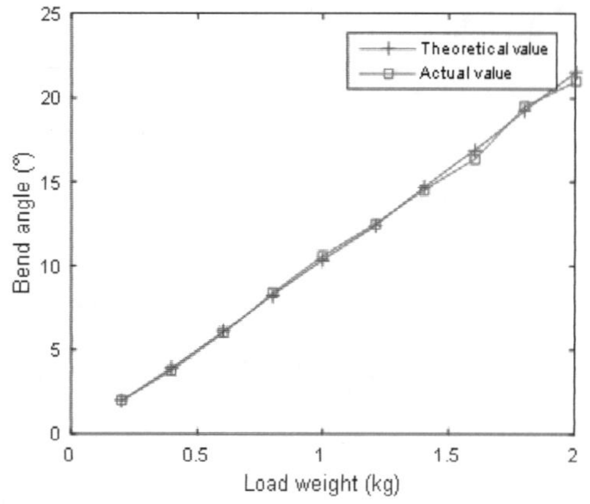

Figure 8. Bending angel under one-direction load.

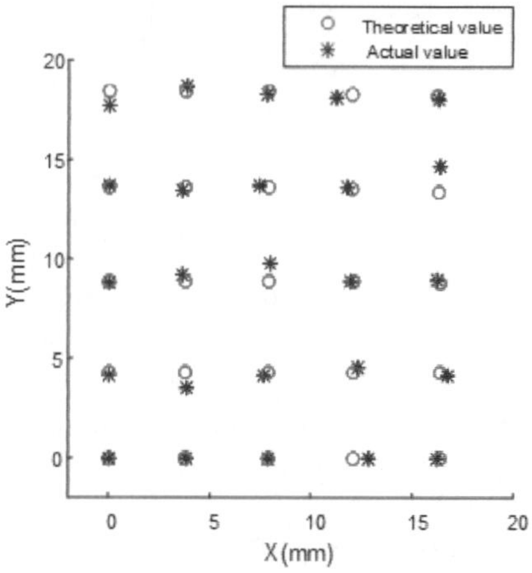

Figure 9. Deformation displacement under two-direction load.

6. Conclusion

The tendon-driven continuum manipulator with crossed notches proposed in this paper is tractable, small, light, safe, and dexterous which can be used in SPSR. The mechanics model describes the manipulator as a variable curvature tube, while the kinematics model considers the manipulator as an equivalent series robot. Both of two models are relatively accurate when depict mechanical and kinematic properties characteristics respectively, and verified experimentally. Future study should focus on increasing accuracy and optimization. Inverse kinematics, force feedback and other issue also need to be considered in the future.

7. Acknowledgments

In this paper, the research was supported by the National Nature Science Foundation of China under Grant No. 61673139.

8. Reference

[1] Davies B 2000 *P I MECH ENG H* **214** 129-40
[2] Gettman M T *et al.* 2011 *EUR UROL* **59** 231-4
[3] Wortman T D, Mondry J M., Farritor S M and Oleynikov D 2013 *IEEE T BIO-MED ENG* **60** 926-29
[4] Petroni G, Niccolini M, Caccavaro S, Quaglia C, Menciassi A, Schostek S, Basili G, Goletti O, Schurr M O and Dario P 2013 *SURG ENDOSC* **27** 1932-7
[5] Quaglia C, Petroni G, Niccolini M, Caccavaro S, Dario P and Menciassi A 2014 *J MECH DESIGN* **136** 105001
[6] Webster R J, Okamura A M and Cowan N J 2007 *IEEE/RSJ Int. Conf. on Intelligent Robots and Systems(Beijing)* pp2857-63
[7] Liao W, Shuang S, Wu K, Lim C M and Ren H 2017 *MED BIOL ENG COMPUT* **55** 403-17
[8] Xu K and Simaan N 2008 *IEEE T ROBOT* **24** 576-87
[9] Li Z and Du R 2013 *INT J ADV ROBOT SYST* **10** 209-20
[10] Z Du, W Yang and W Dong 2015 *J MECH ROBOT* **7** 041017
[11] Gao A, Murphy R, Liu H, Iordachita I. and Armand M 2016 *IEEE-ASME T MECH* **22** 465-75
[12] Lobontiu N 2002 *Compliant mechanisms: design of flexure hinges* CRC press chapter 1 pp 1–16
[13] WU Y and ZHOU Z 2002 *ENG MECH* **6** 026

IOP Conf. Series: Materials Science and Engineering **320** (2018) 012006 doi:10.1088/1757-899X/320/1/012006

A Two-stage Improvement Method for Robot Based 3D Surface Scanning

F B He[1], Y D Liang[1,2], R F Wang[1] and Y S Lin[3]
[1] School of Mechanical Engineering, Dalian University of Technology, Dalian 116024, China
[2] Engineering Training Center, Dalian University of Technology, Dalian 116024, China
[3] Department of Computer Science, Dalian Ocean University, Dalian 116023, China

E-mail: hefuben@mail.dlut.edu.cn

Abstract. As known that the surface of unknown object was difficult to measure or recognize precisely, hence the 3D laser scanning technology was introduced and used properly in surface reconstruction. Usually, the surface scanning speed was slower and the scanning quality would be better, while the speed was faster and the quality would be worse. In this case, the paper presented a new two-stage scanning method in order to pursuit the quality of surface scanning in a faster speed. The first stage was rough scanning to get general point cloud data of object's surface, and then the second stage was specific scanning to repair missing regions which were determined by chord length discrete method. Meanwhile, a system containing a robotic manipulator and a handy scanner was also developed to implement the two-stage scanning method, and relevant paths were planned according to minimum enclosing ball and regional coverage theories.

1. Introduction

Aiming at measuring and recognizing the topography and contour of unknown object, the laser scanning was applied to collect 3D data as much as possible trying to reconstruct the surface. As a non-contact measurement technology, it owned many advantages, which had been widely applied in various engineering fields. To perform automatic 3D laser scanning application, recent studies mainly adopted the method of prediction estimation to plan the sampling points, the viewpoints and the scanning paths so as to obtain the complete 3D information of the surface. For example, to apply a mass vector chain to determine the exploration direction and the accurate position in the viewpoint planning of 3D model reconstruction [1], to present a next-best-scan (NBS) approach to plan collision free path in order to completely scan unknown objects [2], an adaptive sampling method presented to plan the measuring path according to the curvature feature of the fitted B-splines [3], using a model-based viewpoint planning method to inspect freeform surface [4], and a new next-best-view (NBV) planning method proposed for reconstruction of unknown 3D model based on limit visual surface [10], etc. However, the missing region determine errors, the sampling point arrangement errors, the viewpoint deviations and other problems still existed in unknown surface scanning in spite of these studies had been demonstrated in some experimental applications. Furthermore, some of scanning methods might consume amount of resources by algorithm operations in the process and the speed was quite slow in pursuit of the completeness of scanning. To improve the quality of 3D scanning with consideration of speed, the paper presented a two-stage scanning method including the firstly rough scanning and secondly specific scanning. The idea was to scan the object in a faster speed so as to obtain general 3D information of surface, which would reduce scanning time immensely, then to analyse and to determine whether the collected data existed missing regions. If the missing regions were confirmed, one more detailed scanning seemed to be very necessary to complete the scanning process. In the laser 3D scanning, point cloud data was collected from the surface and was able to be converted to spatial data points. Then it needed to apply mathematics algorithms to confirm the locations and orientation of the missing regions. In the scanning path planning, the minimum enclosing ball (MEB) theory [5-8] was introduced, since MEB had been already applied in solving large data sets problem [5,8] and evaluating the barycentre error in large blade optimization process [6,7]. It represented a ball field with the minimum radius to enclose a set of spatial points availably, whose characteristics were suitable to solve the missing problem of 3D surface scanning and to make effective estimation of the area of missing regions. Combining 3D scanning constraints mainly based on limited dimensions in Euclidean space [11,12] and the visibility cone concept which was simply described as scanning vector visual area [13,14], the sampling points and directions of scanning were determined, so that the specific scanning path was planned through computed points and directions. To implement this improvement scanning method, it required the assistance of an automatic platform in order to execute the operations, so that a robotic manipulator mounted with a handy 3D scanner was designed and applied, since its movement trajectories were accurate and the performances of scanning paths were also guaranteed.

2. Rough scanning

The general topography and contour information of object surface were quickly obtained by rough scanning. Through determining the information collected by 3D scanner and analysing the point cloud data of boundaries, it was able to provide necessary computational data support to path planning in specific scanning. From the perspective of discretization, transforming the 3D information of measured surface into point cloud data, the missing scanned regions were determined by the discretization algorithm [9]. As depicted in Figure 1, the original point cloud data was divided into several layers based on the values along Z axis and the data of each layer was projected onto the XOY plane. According to the accuracy of 3D scanner, the distance threshold of projected point cloud was defined by T. It also defined that adjacent projection points on the projection plane were $P_i(x_i, y_i)$ and $P_{i+1}(x_{i+1}, y_{i+1})$ respectively. If the Euclidean norm from P_i to P_{i+1} satisfied the equation (1), it was determined that there was a missing region between P_i and P_{i+1}, that was, 3D information was missing in this region. Meanwhile, P_i and P_{i+1} represented the boundary points of the missing region.

$$\|P_i - P_{i+1}\| = L_i > T \tag{1}$$

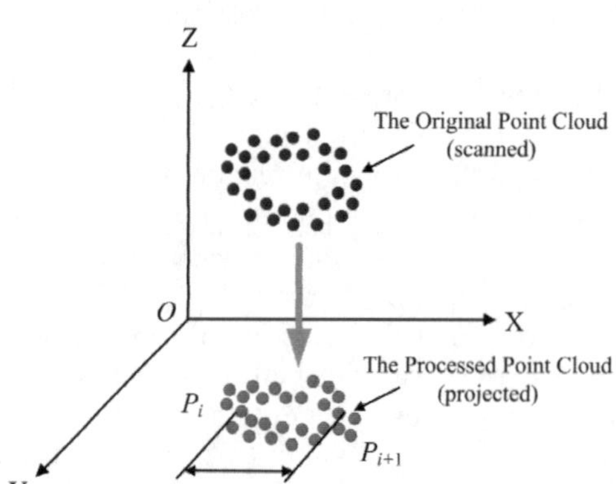

Figure 1. The missing regions determined by point cloud data.

There also existed invalid scanning region which contained incomplete data in surface gridding process, reading and processing the 3D information of the triangular patches included in the region, the estimated value of the normal direction of missing region $\boldsymbol{N}_m$ was expressed by the average value of the normal directions of triangular facets $\boldsymbol{N}_j$ as

$$\boldsymbol{N}_m = \frac{\sum_{j=1}^{m} \boldsymbol{N}_j}{\left|\sum_{j=1}^{m} \boldsymbol{N}_j\right|} \qquad j = 1,2,\cdots,m \tag{2}$$

3. Specific scanning

In the specific scanning process, firstly sampling points of scanning was arranged in order to confirm the repair scanning centre and then scanning paths were planned

3.1 Sampling point arrangement

The MEB theory and the shrinkage optimization algorithm were introduced to enclosing compute the boundary point cloud of the missing regions. The MEB of missing region B_r was obtained with the centre $C_0(x_0, y_0, z_0)$ and the radius R_0 as depicted in Figure 2(a). The ball centre C_0 was also treated as the geometric centre of missing region. Since the estimated value of the normal direction of missing region was $\boldsymbol{N}_m$, the plane Π was defined through the ball centre C_0 and perpendicular to the normal direction $\boldsymbol{N}_m$, so that the sampling points should be on plane Π. Defining arbitrary point $C(x, y, z)$ on plane Π, then $C_0C(x\text{-}x_0, y\text{-}y_0, z\text{-}z_0)$, the plane Π was expressed by equation (3).

$$\boldsymbol{N}_m \bullet \boldsymbol{C}_0\boldsymbol{C} = \boldsymbol{0} \tag{3}$$

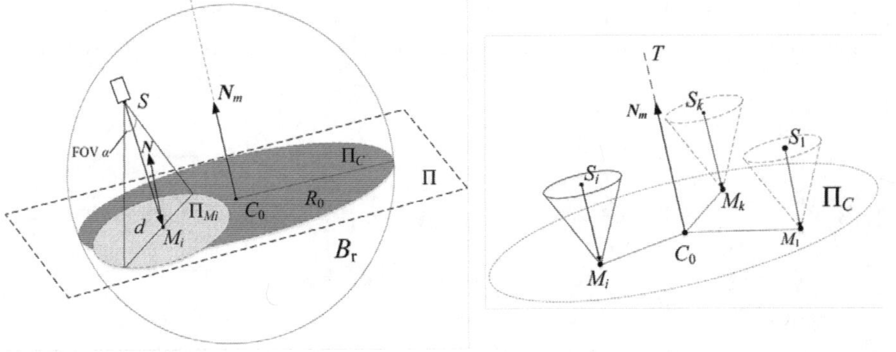

(a) (b)

Figure 2. Specific scanning planning based on MEB.

The cross section of the plane Π and the MEB B_r was the circle Π_C with the centre C_0 and radius R_0, as depicted in Figure 2(a) in blue. Making the sampling point M_i of missing region being on the axis of the field of vision (FOV), the intersection of the set of FOVs corresponding to the viewpoint S and the plane Π formed the circle of vision (COV) Π_{Mi} with the centre M_i and the diameter d, as depicted in Figure 2(a) in grey. To ensure that all positions of missing regions of 3D surface were collected, the problem was converted to compute the minimum number of COVs Π_{Mi} required to cover the circle Π_C. Considering that the missing region was usually small, so that the difference between the radius of the circle Π_{Mi} and the radius of the circle Π_C was little, therefore, the minimum number of FOVs required to cover the MEB's cross section was computed by circumference division method and the spatial coordinates of the sampling points were arranged. Taking the diameter d of COV Π_{Mi} as the chord length of circle Π_C to equally divide the circumference of the circle Π_C, defining that φ was the central angle corresponding to the string d and x was the number of required strings, the relationship between x and φ was expressed by $x=[2\pi/\varphi+1]$, where [.] represented to be rounded down.
According to the circumference division method, the number of strings x depends on the relationship between d and R_0. When $x>5$, the missing region was relatively larger, which rarely appeared, so only

the situation when $x<5$ was considered. When $x=1$, the sampling point M_i was the centre of the circle Π_C, when $x=3$ or $x=4$, the number of sampling points was equal to the number of strings, and the sampling point M_i was located at the midpoint of each chord of equally divided circumference. Since the missing region was quite small in most cases, it was usually required to arrange one sampling point.

3.2 Scanning path planning

On the theoretical basis of the 3D scanning constraints and the visibility cone concept, the scanning viewpoints were arranged and the scanning directions were confirmed, as depicted in Figure 2(b). In order to ensure that each position of the missing region was collected, the viewpoints S_i were arranged on the axis parallel to the normal direction N_m through the sampling point M_i. The height of the viewpoint S_i from the sampling point M_i was between the near depth and the far depth of scanning vision and the scanning direction was C_0N_m. In specific scanning process, path planning was actually to determine the order of each scanning viewpoint, and then to confirm the corresponding movement trajectory of the scanner through the spatial position of each viewpoint. The procedure of scan path planning was described as following:

- To determine the order of scanning viewpoints: connecting each viewpoint in the clockwise direction and scanning the viewpoint on the axis C_0T in the end.
- To plan the scanning path of adjacent viewpoints: applying the cubic spline interpolation method [15] to plan the trajectories between the two adjacent S_{i-1} and S_i so as to ensure the stability of the whole scanning process.
- To confirm the movement trajectory of scanner: establishing the relative spatial position relationship between the object and the scanner, transforming the planned path of viewpoints into the real trajectory of scanner, to complete the 3D scanning and surface reconstruction.

4. Robot based scanning system

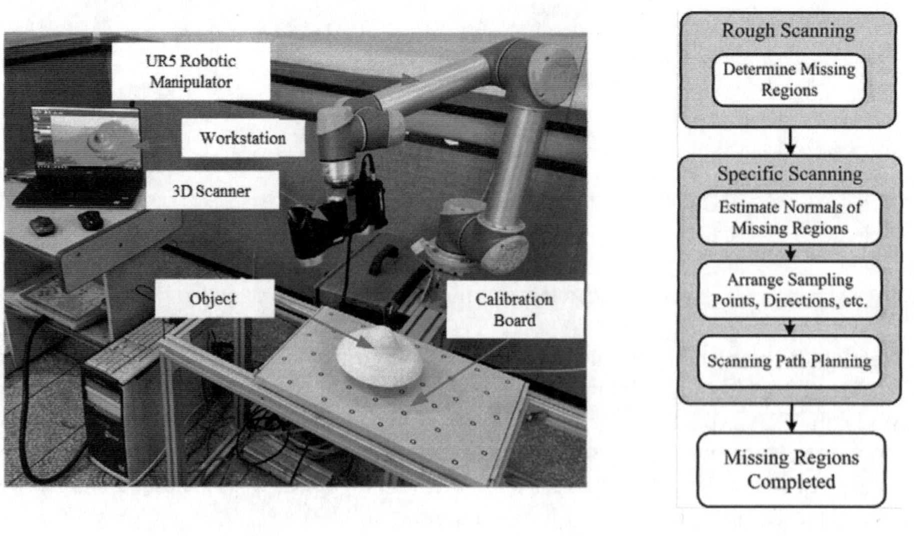

(a) (b)

Figure 3. Design and workflow of robot based scanning system.

In order to verify the effects of the two-stage scanning method, an experimental scanning system was set up as depicted in Figure 3(a). It consisted of a HandyScan300 3D scanner, a UR5 robotic manipulator, a Dell graphics processing station, a calibration board and other components. The main parameters of 3D scanner: the vision distance of was 300mm, the depth of field was 250mm, and the coverage area was 225mm x 250mm at the vision height. Since UR5 robot has six degrees of freedom, strong load capacity and wide work range, it was applied with a customized gripper mounted on the end to carry the 3D scanner in the scanning process. The workflow of this robot based scanning system of the whole robot was depicted in Figure 3(b), and it was also applied with the two-stage scanning method.

In the rough scanning, it required that to scan the surface of object at a faster speed to obtain the general point cloud data, which was able to increase the scanning efficiency with no need for high requirements of 3D information integrity. Similar to the motions of rough machining, the robot carried the scanner to perform alternating movements. In the process, the boresight axis of scanner was always vertically down, the scanning route interval was set to 150mm, and the speed of scanning was set to 0.3m/s, which was faster than the scanning speed by holding the scanner with hands. To roughly scan the object depicted in Figure 3(a), after processing 3D information collected by the scanner, relevant point cloud data was obtained, in which the total number of points was 1.31657E+5 and the superficial area after fitting was 3.858923E+4mm^2. It was easily determined that the missing regions are M1 and M2 based on initial scanning results as depicted in Figure 4(a).

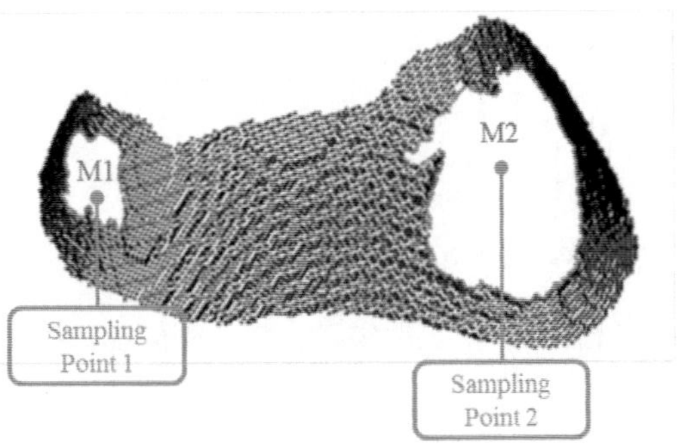

(a)

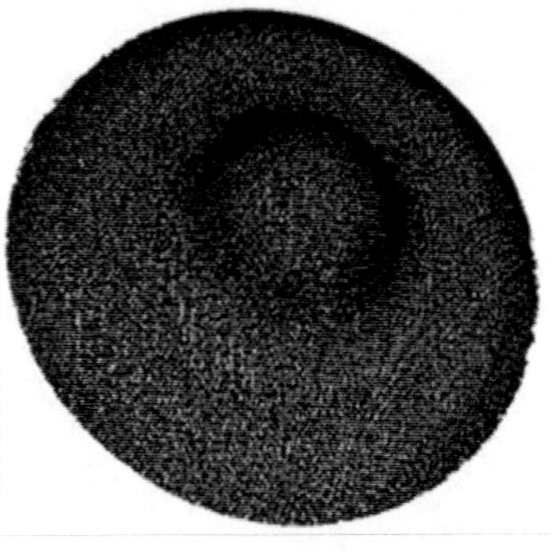

(b)

Figure 4. The process of 3D surface scanning.

Applying the shrinkage optimization method to enclosing compute the point cloud of boundaries of missing regions M1 and M2 respectively, the MEB centre of M1 was obtained at (72.86, -226.61, 188.22) and the radius was 8.12mm, similarly the MEB centre of M2 was at (23.35, -236.77, 198.62) and the radius was 16.07mm. Considering the geometric relationships in vision distance, depth of field, coverage area and other parameters of scanner, the radius of COV was computed to be 93.75mm. The radiuses of circles produced from the MEBs of M1 and M2 intersecting with respective Π planes were 8.12mm and 16.07mm respectively, and the ratios to the radius of COV were 0.09 and 0.17 respectively. The number of sampling points was set to 1 by the circumference division method. Therefore, the sampling points were arranged and the scanning viewpoints, the scanning directions and the scanning paths were all able to be planned.

Then transforming the scan path into the movement trajectories of robot, guided the robot equipped with 3D scanner was guided to repair scanning the missing regions through specific sampling points and scanning directions, the complete point cloud data of object as depicted in Figure 4(b). After that, the number of points included in scanned point cloud data of object became 1.34053E+5 and superficial area after fitting became 3.93980E+4mm^2, increased by 2396 and 808.77mm^2 respectively, compared to the obtained data in rough scanning, which meant that the quality of scanning was improved and the topography of object surface was completed in the end.

5. Conclusion

In order to pursuit the quality of scanning in a faster speed, a new two-stage method was presented in 3D laser surface scanning process for unknown object. In the first rough scanning, the general contour and topography of surface were obtained as point cloud data, and the missing regions were determined by using chord length method. In the second specific scanning, focusing on the missing regions, the exact sampling locations and the scanning paths were planned by MEB theory and regional coverage theory. The improvement method was implemented in a robot based scanning system which contained a robotic manipulator and a handy scanner, with the advantages of convenient operability and fast movability. Utilizing a robotic platform, it not only enhanced the stability of scanning, but also reduced the errors caused by shock and vibration in the scanner's movements. Finally, relevant experiments were carried out and the results were evaluated for the feasibility and effectiveness of this two-stage scanning method, and both of speed and quality of scanning were improved.

6. Acknowledgments

The authors would like to acknowledge that this study was partly supported by National Natural Science Foundation of China (Grant No. 61603067) and Liaoning Province Natural Science Foundation (Grant No. 2015020068).

7. References

[1] Li Y F, He B and Bao P 2005 *Sensors and Actuators A: Physical* **22** 335-44
[2] Kriegel S, Rink C, Bodenmüller T, Narr A, Suppa M and Hirzinger G 2012 *Int. Conf. on Intelligent Robots and Systems* (Vilamoura, Portugal) 2850-56
[3] Lu K Q, Wang W and Chen Z C 2010 *J. Mechanical Engineering* **46** 143-49
[4] Wu Q, Zou W and Xu D 2016 *Int. J. Automation and Computing* **13** 42-52
[5] Cervantes J, Li X, Yu W and Li K 2008 *Neurocomputing* **71** 611-19
[6] Liang Y D, Sun J F, He F B and Lu Y H 2015 *China Mechanical Engineering* **26** 756-61
[7] He F B, Liang Y D, Sun J F, Wang Z X and Guo C 2015 *J. Mechanical Engineering* **51** 64-70
[8] Song Q J, Xiao X M, Jiang H Y and Zhao X G 2015 *J. Mechanical Science and Technology* **29** 3467-73
[9] Wu B H and Wang S J 2004 *Mechanical Science and Technology* **23** 1363-65
[10] Zhou X L, He B W and Li Y F 2009 *Prod. Int. Conf. on Robotics and Biomimetics* (Guilin, China) p 965-70
[11] Kawasaki H and Furukawa R 2007 *Prod. 6th. Int. Conf. on 3-D Digital Imaging and Modeling* (Montreal, QC, Canada) p149-58
[12] Ozan Ş and Gümüştekin Ş 2014 *Digital Signal Processing* **24** 231-43
[13] Lee K H and Park H P 2000 *Robotics and Computer-Integrated Manufacturing* **16** 201-10.
[14] Lee R S, Lin Y H, Tseng M Y and Wu W S 2010 *Int. J. Computer Integrated Manufacturing* **23** 630-39
[15] Gasparetto A and Zanotto V 2010 *Advances in Engineering Software* **41** 548-56

IOP Conf. Series: Materials Science and Engineering **320** (2018) 012007 doi:10.1088/1757-899X/320/1/012007

Eye-in-Hand Manipulation for Remote Handling: Experimental Setup

Longchuan Niu[1], Olli Suominen[2], Mohammad M. Aref[1], Jouni Mattila[1], Emilio Ruiz[3] and Salvador Esque[3]

[1] Laboratory of Automation and Hydraulics Engineering, Tampere University of Technology, FI-33101, Tampere, Finland

[2] Laboratory of Signal Processing, Tampere University of Technology, FI-33101, Tampere, Finland

[3] Fusion for Energy, Barcelona, Spain

Abstract. A prototype for eye-in-hand manipulation in the context of remote handling in the International Thermonuclear Experimental Reactor (ITER)[1] is presented in this paper. The setup consists of an industrial robot manipulator with a modified open control architecture and equipped with a pair of stereoscopic cameras, a force/torque sensor, and pneumatic tools. It is controlled through a haptic device in a mock-up environment. The industrial robot controller has been replaced by a single industrial PC running Xenomai that has a real-time connection to both the robot controller and another Linux PC running as the controller for the haptic device. The new remote handling control environment enables further development of advanced control schemes for autonomous and semi-autonomous manipulation tasks. This setup benefits from a stereovision system for accurate tracking of the target objects with irregular shapes. The overall environmental setup successfully demonstrates the required robustness and precision that remote handling tasks need.

1. Introduction

Remote handling (RH) is vitally important in the maintenance and operation of the ITER nuclear reactor where human access is impossible. RH tasks are performed indirectly with remotely operated robotic manipulators using man-in-the-loop teleoperation. Vision and force display devices are essential for any teleoperation system to assist operators in manipulation tasks in inaccessible environments [1]. Without haptic feedback [2], peg-in-hole tasks are very difficult to complete when the clearances are very small. Precise and reliable teleoperation could be achieved using marker-based advanced imaging methods, but in the challenging ITER environment, which is under high doses of radiation, any marker object material may suffer from erosion and deformation. Previous work on Computer Aided Teleoperation (CAT) based on 2D template matching and a fixed camera discovered significant limitations of the monocular approach [3]. As a practical solution to cover the above design challenges, in our environment setup, integration of eye-in-hand stereoscopic cameras into the control

[1] The nuclear-fusion reactor, constructed within the ITER project (http://www.iter.org).

chain with the specifically developed Iterative Closest Point (ICP) algorithm [4] provides robust and accurate pose estimation and object tracking. This paper is organized as follows. In the next section, the architecture of the experimental setup is introduced by its hardware and software subsystems. Then, section 3 explains the challenges and solutions for integration of stereoscopic visual feedback in such an architecture. Finally, section 4 shows the system performance in practice.

2. System Architecture

Figure 1. Remote Handling Mock-up Scene.

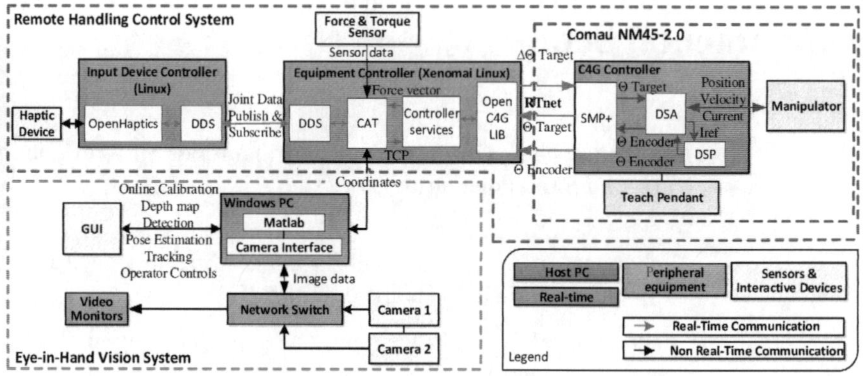

Figure 2. System architecture.

An experimental setup of the remote handling control system (RHCS) for teleoperation of the Comau SMART NM45-2.0 industrial robot is shown in Figure 1. The system architecture, in Figure 2, comprises the eye-in-hand vision system and RHCS.

2.1. Operating and Development Environments

In the RH architecture, there are a variety of operation environments. Windows environments are flexible and good for user interfaces, however, they do not have characteristics of a real-time operation system, such as determinism. Considering Linux-based systems, Xenomai [5] is a free, POSIX-compatible, real-time kernel extension framework that is easier for programming industrial applications. Similarly, another popular real-time Linux-based system is the Real-Time Application Interface (RTAI) [6]. The performance of both running as dual-kernel schemes is nearly identical [7], provided both use RTnet [8], which is a hard real-time networking framework. The Comau C4G controller provides a software extension called C4G open, which allows for interacting with robots via an external PC (in our case, a Debian Linux PC enhanced by Xenomai) by means of the RTnet protocol stack (Figure 2). The input device controller is another Debian Linux PC installed with OpenHaptics to support haptic devices and OpenSplice data distributroin service (DDS) for real-time communication with the Xenomai Linux PC. The host PC for the vision system is a Windows 10 system installed with Matlab and the camera interface for camera access. As explained above, this kind of setup can face heterogeneous subsystems because of the different requirements for each part of the system, especially when it has to be integrated or coordinated with the pre-existing industrial setups.

2.2. Remote Handling Control System

The RHCS is a master-slave structure, where the master input device is the haptic device and the robotic manipulator Comau NM45-2.0 is the slave. The equipment controller connects them in the middle, together with force sensor and the Eye-in-Hand vision system. The Xenomai Linux PC [9] mentioned previously was developed to become an equipment controller featuring the following functionalities: real-time communication with the Comau C4G controllers, Comau NM45-2.0 low-level control, trajectory generation, Comau NM45-2.0 kinematics, publish and subscribe routines for the robotic joint position data, force sensors reading, and vision data reading and sending. The RH operator uses the eye-in-hand vision application through the Graphical User Interface (GUI) and bilaterally teleoperates the Comau NM45-2.0 manipulator through Phantom Omni, a six-Degree-of-Freedom (DOF) haptic display. The design of the bilateral teleoperation control for the input device controller follows Hooke's law as the virtual impedance:

$$\boldsymbol{F} = \begin{cases} \mathbf{0} & f(\boldsymbol{x}) > f(\boldsymbol{x_0}) \\ \boldsymbol{Kx} + \boldsymbol{B\dot{x}} & f(\boldsymbol{x}) \leq f(\boldsymbol{x_0}) \end{cases}$$

where F is the contact force, K is stiffness constant value, B is the damping coefficient along K, $\boldsymbol{x}$ is the impedance type of penetration vector and $\boldsymbol{x_0}$ is a given point in the vicinity of the obstacle. The input device controller is implemented in this impedance-based position-force approach to prevent undesirable movement changes.

2.3. Communication Interfaces

This heterogeneous system requires seamless, time-critical communication. To ensure real-time deterministic communication between the equipment controller and the Comau C4G controller, the RTnet is used to carry the controller specific real-time queues with a 2 ms communication cycle time. Communication between the input device controller and equipment controller was implemented using the OpenSplice DDS middleware at a 1-kHz frequency. Communication between the 3D node and the RHCS is with UDP packets at 1Hz.

3. Integration of Eye-in-Hand Vision System in RH, the 3D Node

The main aim of our project is to discover the potential of stereoscopic imaging for remote handling. The setup of the stereo cameras was done according to the topology design, which takes view angle, stereoscopic precision, and lighting into account. Figure 3 describes the processing stages of the 3D Node module, i.e., the eye-in-hand vision system.

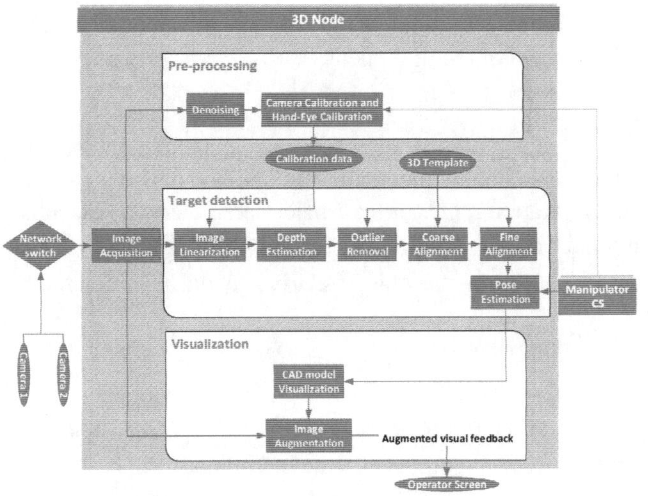

Figure 3. Flowchart of the 3D Node.

IOP Conf. Series: Materials Science and Engineering **320** (2018) 012007 doi:10.1088/1757-899X/320/1/012007

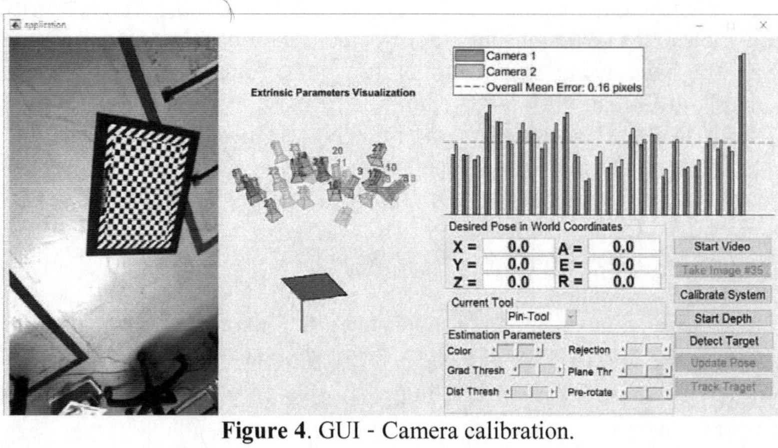

Figure 4. GUI - Camera calibration.

3.1. Implementation of 3D Node

Two GenTL-compliant cameras (Prosilica GE1900) are synchronized using the OUT pin of camera 1 connected with the LINE1 pin of camera 2, automatically triggering both cameras once camera 1 is triggered through the software. This ensures both images have the same content during camera motion. The 3D node acquires images from the stereo cameras through GigE. A setup stage is done outside the operation environment, where the cameras are calibrated using test images of a special calibration pattern (Figure 4). These images are used to extract the intrinsic and extrinsic parameters [10] of the cameras. Meanwhile, the calibration module also performs the hand-eye calibration, referencing the camera's optical center to the tool center point (TCP) of the manipulator. Based on the camera calibration information, the pre-processing stage handles denoising removal of optical distortions (Figure 3). The 3D reconstruction module performs depth estimation (stereo matching) and outlier detection to achieve a clean point cloud of the target scene. This is used for a coarse alignment of the target. The fine alignment step refines the pose estimate by comparing the acquired depth image with the one generated from the known CAD model. This alignment provides the tracking and positioning capabilities while the system is in use. The visualization stage generates the 3D Node GUI with all the user controls and views. In addition, it provides the augmented camera images to the user screen and a visual confirmation of the success of the alignment.

3.2. Mapping of Coordinates

The object pose in camera coordinates has to be transformed into the robot world coordinates, for which the eye-in-hand calibration [11] of extrinsic parameters is essential. With the estimated transform between the model object and the one sensed with the depth-estimation procedure (M_{icp}), one can finally reconstruct a pose of the object relative to the manipulator flange and to the world coordinates (P_{obj}).

$$P_{obj} = R_{robot} X_{handeye} M_{icp}^{-1},$$

where R_{robot} is a pose of a manipulator flange (TCP) and $X_{handeye}$ is the hand-eye matrix, transforming from the camera coordinate space to the manipulator TCP.

The point of interest in the object model, I_{poi}, (i.e., the point where the tool should be inserted), which is manually pre-calibrated in advance, can be re-projected in the camera space using the found fine alignment. Then, the location of this point relative to the manipulator TCP can be recovered by using the hand-eye relationship matrix, recovered during the eye-in-hand calibration.

$$D_{desired} = T_{tool} X_{handeye}^{-1} R_{robot}^{-1} P_{obj}^{-1}\ I_{poi},$$

where $D_{desired}$ is the desired location and orientation of the tool prior to insertion (in the world coordinate system), T_{tool} is the tool transformation matrix. The R_{robot} matrix is estimated online using the TCP values from the manipulator.

4. Experimental Results

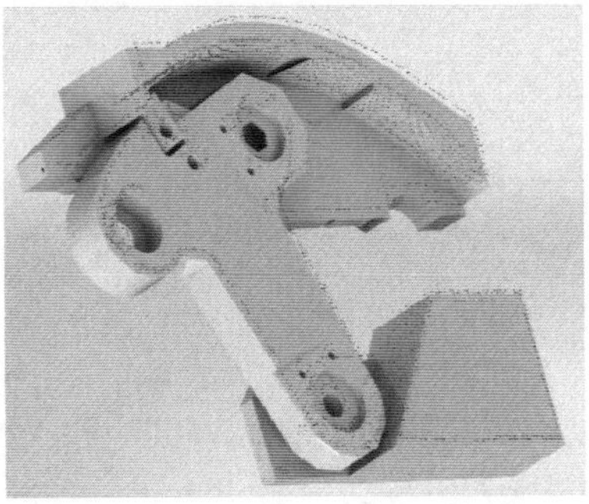

Figure 5. CAD model compared to detected surfaces in the measured point cloud.

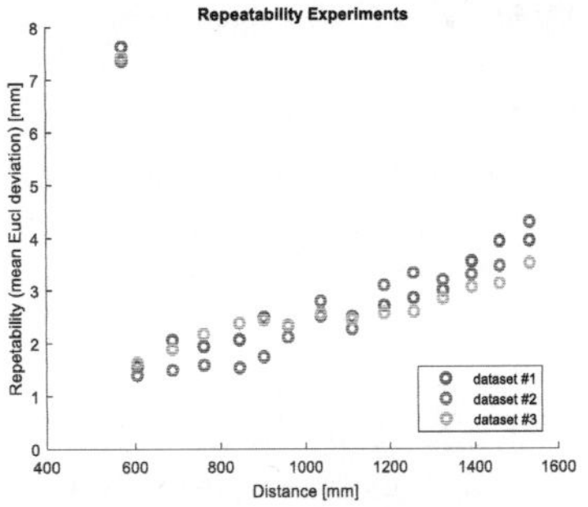

Figure 6. Repeatability (mean deviation) in respect to depth (z-distance).

As shown in Figure 5, the target object knuckle can be complicated such that some parts of it can be occluded. Based on the acquired cloud of points (Figure 5, red points), after filtering out the unnecessary parts, it is possible to match with the existing CAD model. This way, the pose of the object is estimated with more accuracy, making it feasible to use its information in the real-time visualization and feedback of the manipulation loop.

One of the important quality aspects for the pose estimation system is its repeatability. That is, a good system can re-measure the same position, taking into account that measurements are independent. We used three sets of experiments. Each of those stereo-images of the knuckle object were taken at different distances ranging from 60 cm to 160 cm. For every experiment, we used a slightly varying orientation (i.e., relative rotation) of a knuckle object, which was approximately the same within all images within the experiment. By measuring the Euclidean distance between the centroid and every other estimate, one can obtain the deviation for a particular stereo-pair. While taking depth, using z-value as a reference variable, we can plot the figure where the horizontal axis is depth (z-distance between camera and the knuckle) and the vertical axis shows the respective deviation values. In Figure 6, when being too close to the target object, camera can no longer observe all the distinctive edges, thus deteriorating overall performance. Within the main operating range, all datasets present good and consistent results. As distance increases, the repeatability error grows but also becomes unstable, which could indicate the existence of an upper limit for the system.

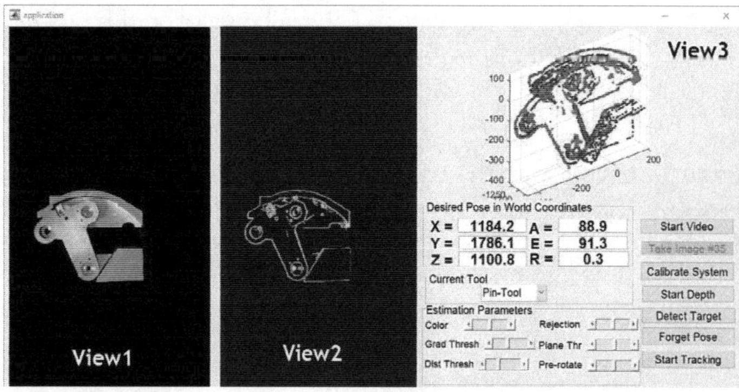

(a) matching (b) depth (c) an estimated pose

Figure 7. (a) the Augmented knuckle completely matched to the real knuckle (b) depth map (c) The point clouds of the knuckle from the CAD model in blue and from the sensed one in red.

(a) the pin tool (b) the jack tool

Figure 8. The scene of the peg-in-hole insertion for the pin tool and the jack tool, where the clearance is only 5mm between the jack tool and the knuckle.

The 3D Node GUI receives user input and provides visual feedback and pose estimation data of the target object, the knuckle. The captured image in Figure 7(a) is augmented with the image of the rendered knuckle at the estimated position. As the captured and rendered images are grayscale, we use color mapping to show two images simultaneously on the same plane. The user can directly see how well images were aligned and verify when the pose estimation was incorrect. This way, even a small angular error can be more noticeable by the user. The 3D node continually processes the live data and displays the estimated pose of the knuckle in the robot manipulator world coordinate, as shown in Figure 7(c), The sampled point clouds of the knuckle from CAD model in blue and the surface point clouds sensed from the real scene in red are shown in view 3, the pose of the detected knuckle is shown at the same time, from which the user can operate the manipulator for a "peg-in-hole" assembly task, as illustrated in Figure 8. The comparative augmented visualization here, significantly improves tool insertion in accuracy and speed.

5. Conclusion

In the current setup environment with the aid of eye-in-hand stereoscopic visual feedback, a RH operator can use the haptic device to accomplish required use cases, such as pickup and drop-off of different tools, and perform peg-in-hole insertion tasks where clearance can be only 5mm. The prototype system has shown to not only increase the speed, usability, and safety of remote handling operation, but also helps the operator to reduce stress. Overall, the experimental setup has demonstrated great reliability, and the system's precision meets the needs set by the unique ITER operating environment.

6. References

[1] M. Panzirsch, R. Balachandran, J. Artigas, C. Riecke, M. Ferre and A. Albu-Schaeffer, "Haptic intention augmentation for cooperative teleoperation," 2017 IEEE International Conference on Robotics and Automation (ICRA), Singapore, 2017, pp. 5335-5341.

[2] J. Kofman, Xianghai Wu, T. J. Luu and S. Verma, "Teleoperation of a robot manipulator using a vision-based human-robot interface," in *IEEE Transactions on Industrial Electronics*, vol. 52, no. 5, pp. 1206-1219, Oct. 2005.

[3] Z. Ziaei, A. Hahto, J. Mattila, M. Siuko, and L. Semeraro, “Real-time markerless augmented reality for remote handling system in bad viewing conditions,” Fusion Engineering and Design, vol. 86, no. 9, pp. 2033–2038, 2011.

[4] Rusinkiewicz S, Levoy M. Efficient variants of the ICP algorithm. In3-D Digital Imaging and Modeling, 2001. Proceedings. Third International Conference on 2001 (pp. 145-152). IEEE.

[5] Gerum P. Xenomai-Implementing a RTOS emulation framework on GNU/Linux. White Paper, Xenomai. 2004 Apr:1-2.

[6] Mantegazza P, Dozio EL, Papacharalambous S. RTAI: Real time application interface. Linux Journal. 2000 Apr 1;2000(72es):10.

[7] A. Barbalace, A. Luchetta, G. Manduchi, M. Moro, A. Soppelsa and C. Taliercio, "Performance Comparison of VxWorks, Linux, RTAI, and Xenomai in a Hard Real-Time Application," in *IEEE Transactions on Nuclear Science*, vol. 55, no. 1, pp. 435-439, Feb. 2008.

[8] J. Kiszka and B. Wagner, "RTnet - a flexible hard real-time networking framework," *2005 IEEE Conference on Emerging Technologies and Factory Automation*, Catania, 2005, pp. 8 pp.-456.

[9] Alho P., Mattila J. (2013) Real-Time Service-Oriented Architectures: A Data-Centric Implementation for Distributed and Heterogeneous Robotic System. In: Schirner G., Götz M., Rettberg A., Zanella M.C., Rammig F.J. (eds) Embedded Systems: Design, Analysis and Verification. IESS 2013. IFIP Advances in Information and Communication Technology, vol 403. Springer, Berlin, Heidelberg

[10] Forsyth, David, and Jean Ponce, “Computer vision: a modern approach”, Upper Saddle River, NJ; London: Prentice Hall, 2011.

[11] Roger Y Tsai and Reimar K Lenz, “Real time versatile robotics hand/eye calibration using 3d machine vision,” in Robotics and Au-tomation, 1988. Proceedings., 1988 IEEE International Conference on. IEEE, 1988, pp. 554–561.

7. Acknowledgments

The work leading to this publication has been funded in part by Fusion for Energy and TEKES under Grant F4E-GRT-0689. This publication reflects the views only of the authors, and Fusion for Energy or TEKES cannot be held responsible for any use which may be made of the information contained therein.

A novel robotic platform for single-port abdominal surgery

Satwinder Singh[1], Jo L K Cheung[1], Biji Sreedhar[1], Xuyen Dai Hoa[1], Hoi Pang Ng[2] and Chung Kwong Yeung[1]

[1] Bio-Medical Engineering (HK) Limited, Cyberport, Hong Kong
[2]NISI Bio-Medical Engineering (Dongguan) Limited, Dongguan Province, PRC

Abstract. In this paper, a novel robot-assisted platform for single-port minimally invasive surgery is presented. A miniaturized seven degrees of freedom (dof) fully internalized in-vivo actuated robotic arm is designed. Due to in-vivo actuation, the system has a smaller footprint and can generate 20 N of gripping force. The complete work envelop of the robotic arms is 252 mm × 192 mm × 322 m. With the assistance of the cannula-swivel system, the robotic arms can also be re-positioned and have multi-quadrant reachability without any additional incision. Surgical tasks, such as lifting, gripping suturing and knot tying that are commonly used in a standard surgical procedure, were performed to verify the dexterity of the robotic arms. A single-port trans-abdominal cholecystectomy in a porcine model was successfully performed to further validate its functionality.

1. Introduction

In conventional open surgeries, large incisions are often made to allow surgeons full visualization of the targeted surgical site but this is often associated with increased pain, blood loss, possible risk of infection and prolonged hospitalization to the patient [1]. Computer-assisted robotic minimally invasive surgery (MIS) has become increasingly popular over conventional open surgical procedures as it can allow dexterous manoeuvrability of surgical instruments within the body cavity through small incisions [2]. Typical MIS procedures with existing technologies require multiple small incisions or a single incision for the insertion of surgical instruments [3].

In multi-port systems, more than one incision is required to insert the laparoscopic instruments, camera and additional auxiliary surgical tool inside the body. The surgeons acclimatize themselves to operate these systems through training, despite inadequate workspace and poor ergonomics (limited space) that makes certain manipulations difficult [4]. As the motors in these systems are externally driven, re-positioning of the laparoscopic instruments during the surgical procedure is difficult and thus require more incisions. In a more recent development, the instruments for the da Vinci Xi system (Intuitive Surgical, Inc.) can be repositioned, however, this is for multi-port access only [5].

Single-port robot assisted surgical systems is a technological advancement that enhances the existing benefits of MIS. In these systems, only one incision (usually through the umbilicus) is required to reach the surgical site. Despite potential advantages of single-port systems, several challenges still exist such as poor triangulation of surgical instruments, poor working ergonomics for the insertion of multiple instruments into the abdominal cavity and the inability to apply off-axis forces [6].

Recently, various groups have developed single-site surgical robotic systems, such as the da Vinci Sp system from Intuitive Surgicals Inc., SPORT surgical system from Titan Medical Inc., single-port system from Samsung Advanced Institute of Technology [7,8]. In systems that have external-motor driven arms, the system footprint is usually bulky and it is difficult to reposition the arms, however, torque output is higher as compared to compact systems [9].

The concept of developing miniaturized in-vivo robotic instruments with the state-of-the-art technology and endoscopic navigation has become compelling in recent decades [10]. SPRINT, a 6-dof robotic system, has four in-vivo actuators and a shoulder joint operating outside the human body [2]. It consists of two robotic arms (diameter 18 mm). The system is inserted in the abdominal cavity through a cannula of 34 mm diameter. Due to limited in-vivo dof, it is not possible for the surgeon to re-position the robotic arms once they are inserted. In a system designed by Wortman et al., a pair of in-vivo actuated 7-dof robotic arms (diameter 26 mm) has a feature of repositioning by increasing their default incision size (30 mm) [11]. With a smaller incision size requirement, systems possessing snake-like robotic arms are designed to work in a confined workspace, however, torque availability is comparatively less [12,13].

The major challenges in single-port surgical systems' design process are to reduce the incision size, enough torque availability at the end-effector and possibility of arms re-positioning without increasing the incision size. In this work, a new design of fully in-vivo actuated 7-dof robotic arm and its integration with cannula-swivel system are presented. The integration of whole robotic platform provides the possible solution to the challenges discussed above.

2. System design and prototype

The proposed system consists of two serial arms sequentially inserted parallel to the axis of cannula, as shown in figure 1. The diameter and length of the arms is 15.8 mm and 200 mm, respectively. The cannula is a hollow tube with an outer diameter of 20 mm and an inner diameter of 16 mm. It is used as an access port into the abdominal cavity. The robotic arms are deployed to the working position by the embedded motors inside the arms and are securely anchored to the flaps of the cannula by lock pins [14].

The external swivel system (shown in figure 1) can adjust the insertion of the cannula to re-position the robotic arms even when arms are deployed inside the abdominal cavity. This system has three independent dof for telescopic, yaw and pitch adjustments of the cannula. The centre of rotation of the pitch is at the point of insertion to avoid an increase in incision size while re-positioning of the arms. These motor-actuated movements are not considered as additional dofs of the system.

2.1 Robotic arms and master console

The serial arm designed for this project has 7-dof and the Denavit-Hartenberg (D-H) conventions are utilized to attach the co-ordinate frame to each joint as shown in figure 2 [15]. For the first 6-dof of the robotic arm, seven frames are attached to compute six transformation matrices. The 7th dof is for the movement of end effector such as for grasping, cutting etc., therefore, it is not included in the forward kinematics computations. Each matrix establishes a relation between i^{th} and $i\text{-}1^{th}$ frame.

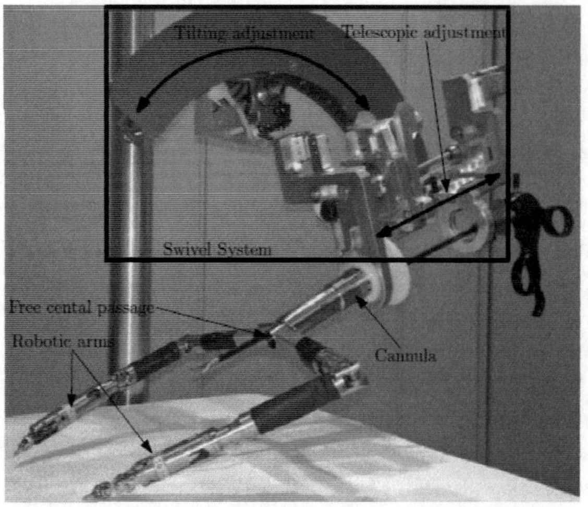

Figure 1. Prototype of the surgical robotic system showing the robotic arms that have been inserted through the cannula and anchored to the external swivel system.

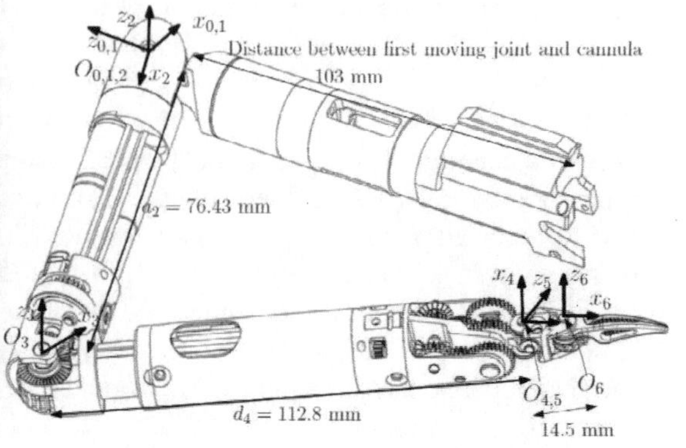

Figure 2. Robotic arm configuration.

Table 1. D-H parameters table for the proposed robotic arm.

Link no.	Link length (a_{i-1})	Twist angle (α_{i-1})	Joint offset (d_i)	Joint angle (θ_i)
1	0	0	0	$-\frac{\pi}{4} \leq \theta_1 \leq \frac{\pi}{4}$
2	0	$\frac{\pi}{2}$	0	$0 \leq \theta_2 \leq \frac{\pi}{2}$
3	a_2	0	0	$0 \leq \theta_3 \leq \frac{7\pi}{12}$
4	0	$-\frac{\pi}{2}$	d_4	$0 \leq \theta_4 \leq \pi$
5	0	$\frac{\pi}{2}$	0	$-\frac{\pi}{2} \leq \theta_5 \leq \frac{\pi}{2}$
6	a_5	$-\frac{\pi}{2}$	0	$-\frac{\pi}{2} \leq \theta_6 \leq \frac{\pi}{2}$

The kinematic equations for the end effector are computed using the homogeneous transformation matrix for the D-H parameters given in table 1. All the joints in the system are revolute, therefore, only joint angles (θ_i) are variables and all the other D-H parameters i.e. link twist (α_{i-1}), link length (a_{i-1}) and joint offset (d_i) are fixed. Utilizing D-H parameters given in the table 1, end-effector's position vector can be computed as,

$$p_x = a_2C_1C_2 - a_5\big(C_5(S_1S_4 - C_1C_2C_3C_4) + S_5(C_1C_2S_3 - C_3S_1 + C_1C_3S_2)\big) - d_4(-C_1C_2S_3 - C_1S_2C_3),$$

$$p_y = a_5\big(C_5(C_1S_4 + C_2C_3C_4S_4) - S_5(C_1C_3 + C_2S_1S_3 + C_3S_1S_2)\big) - d_4(-S_1C_2S_3 - S_1S_2C_3) + a_2S_1S_2,$$

$$p_z = d_4(-S_2S_3 + C_2C_3) + a_2S_2 + a_5(S_5(C_2C_3 - S_2S_3) + C_3C_4C_5S_2),$$

where, p_x, p_y and p_z are the position vectors of the end effector in x, y and z direction, respectively. The *C* and *S* are abbreviation used for cosine and sine functions whereas 1, 2...,5 in the subscript of the *C* and *S* are θ_1, θ_2......, θ_5, respectively.

In the proposed system, a pair of 7-dof robotic arms are tele-operated with the help of two master manipulators. The first 3-dof of master manipulators are responsible for the 3-axis translational movements of corresponding first three joints of the robotic arm. The end-effector movements are controlled by the surgeon's fingers. A foot pedal is provided for re-positioning of the master manipulators to retain the surgeon's working posture. This foot pedal is used to disengage the control and to lock the robotic arms at the concurrent posture. After adjusting the working posture of the master manipulator, clutch is engaged to control the arms at the locked posture.

3. Results and Discussions

Forces and the workspace related measurements are presented in table 2. The maximum measured gripping force (20 N) is comparable to the force reported using the da Vinci Si system [16]. The arms can lift a load of 500 g when the total stretched length (shortest distance measured from first joint to end-effector) is 112.3 mm. The shared work volume of this system (shown in figure 3) is also appropriate as compared to the requirements reported for various robot assisted surgical tasks by Hwang et al. [17]. Shared workspace refers to the operative space where both robotic arms can reach. Currently, the distance between the first actuated joint of the robotic arm from the cannula centre is 103 mm (shown in figure 2) which is required for better triangulation. This distance can be decreased to enhance the shared workspace but at the expense of triangulation. In a single-port system, there is a trade-off between shared workspace and triangulation of the arms.
The proposed overall system design requires a small incision (for the insertion of 20 mm diameter cannula), possesses multi-quadrant access and occupies a small footprint in the operating room. The robotic arms can be re-positioned using the cannula-swivel system combination. The swivel system gives the flexibility of re-positioning of the arms without increasing the size of incision. With this feature, it can allow multi-quadrant access for the surgeon through a single-port approach. Furthermore, there is a requirement of an extra dof to achieve remote centre-of-motion (RCM) in some of the reported systems which is not required in the current design due to all in-vivo actuated joints of robotic arms.

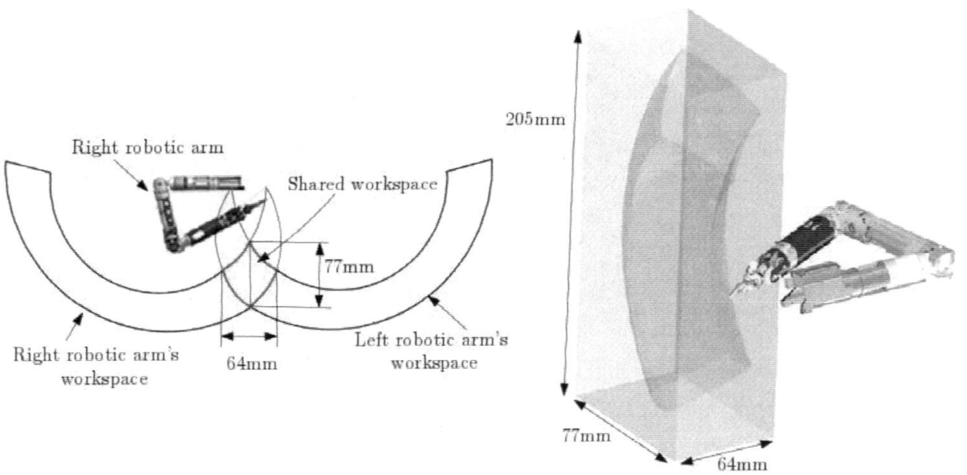

Figure 3. Shared workspace representation of both the robotic arms.

Table 2. Force, power and workspace measurements

Attribute	Measurement
Gripping force	20 N
Payload-lifting	500 g
Output power	469 mW
Total workspace	252 mm × 192 mm × 322 mm
Shared workspace	205 mm × 77 mm × 64 mm

Surgical tasks in laparoscopic surgery, such as, intracorporeal suturing and knot-tying that require high dexterity of the robotic arms, were used to validate the system's performance. Precise movements of all the joints and the end-effector are required to conduct these tasks. Using the C loop technique, the robotic arms demonstrated its ability to tie a knot. As shown in figure 4a, the knot was tied by the two arms pulling in the opposite direction. The feasibility of the robotic arms, in terms of design objectives, was also evaluated in a non-survival porcine model. The animal procedures were performed at the Surgical Skill Centre of The Hong Kong University with experimental protocols approved by the Committee on the Use of Living Animals in Teaching and Research. The robotic arms were fitted with a grasper and a cautery hook that enabled the surgeon to grasp, manipulate and cauterize living tissues. The gall bladder was successfully removed by electrocautery using the robotic arms. Figure 4b shows a view of robotic arms during a technically challenging cholecystectomy procedure. The results have demonstrated that the proposed robotic arm design is able to perform basic surgical tasks required for an abdominal surgery.

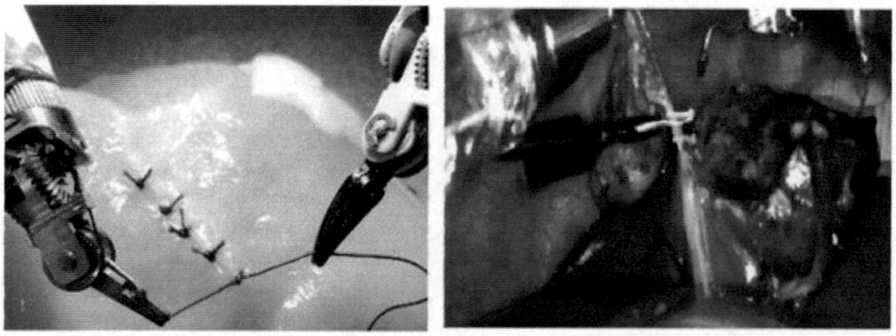

(a) (b)

Figure 4. Surgical tasks (a) Knot tying task (b) Gall bladder removal.

4. Conclusion

The aim of the current work is to present the design and to evaluate a totally in-vivo actuated miniature surgical robot platform for abdominal surgeries. The proposed system is compact (due to fully in-vivo actuation), able to re-position without increasing the incision size and has sufficient torque to perform various surgical tasks inside the abdomen. Results have shown that the robot arm can accomplish the desired surgical tasks with high dexterity. In future work, optimal dimensional analysis is required to enhance the shared workspace.

5. Acknowledgement

This work is funded by the Innovation and Technology Commission of HKSAR (Project Ref. No.: ITS/149/13FX) and Bio-Medical Engineering (HK) Limited.

6. References

[1] Camarillo D B, Krummel T M and Salisbury J K 2004 Robotic technology in surgery: Past, present, and future *Am. J. Surg.* **188** 2–15

[2] Quaglia C, Petroni G, Niccolini M, Caccavaro S, Dario P and Menciassi A 2014 Design of a Compact Robotic Manipulator for Single-Port Laparoscopy *J. Mech. Des.* **136** 105001

[3] Ortmaier T, Weiss H and Falk V 2004 Design requirements for a new robot for minimally invasive surgery *Ind. Robot An Int. J.* **31** 493–8

[4] Kazanzides P, Fichtinger G, Hager G D, Okamura A M, Whitcomb L L and Taylor R H 2008 Surgical and interventional robotics - Core concepts, technology, and design *IEEE Robot. Autom. Mag.* **15** 122–30

[5] Wilson T G 2014 Advancement of technology and its impact on urologists: Release of the da Vinci Xi, A new surgical robot *Eur. Urol.* **66** 793–4

[6] Isaac-Lowry O J, Okamoto S, Pedram S A, Woo R and Berkelman P 2017 Compact teleoperated laparoendoscopic single-site robotic surgical system: Kinematics, control, and operation *Int. J. Med. Robot. Comput. Assist. Surg.* e1811

[7] Beasley R 2012 Medical Robots: Current Systems and Research Directions *J. Robot.* 2012

[8] Choi H, Kwak H, Lim Y and Kim H 2014 Surgical robot for single-incision laparoscopic surgery *IEEE Trans. Biomed. Eng.* **61** 2458–66

[9] Autorino R, Kaouk J H, Stolzenburg J U, Gill I S, Mottrie A, Tewari A and Cadeddu J A 2013 Current status and future directions of robotic single-site surgery: a systematic review *Eur. Urol.* **63** 266–80

[10] Díaz C E, Fernández R, Armada M and García Gutiérrez F de J 2015 State of the art in robots used in minimally invasive surgeries. Natural Orifice Transluminal Surgery (NOTES) as a particular case *Ind. Robot An Int. J.* **42** 508–32

[11] Wortman T D, Meyer A, Dolghi O, Lehman A C, McCormick R L, Farritor S M and Oleynikov D 2012 Miniature surgical robot for laparoendoscopic single-incision colectomy *Surg. Endosc.* **26** 727–31

[12] Mitchell C R, Hendrick R J, Webster R J III and Herrell S D 2016 Toward Improving Transurethral Prostate Surgery: Development and Initial Experiments with a Prototype Concentric Tube Robotic Platform *J. Endourol.* **30** 692–6

[13] Shang J, Leibrandt K, Giataganas P, Vitiello V, Seneci C A, Wisanuvej P, Liu J, Gras G, Clark J, Darzi A and Yang G-Z 2017 A Single-Port Robotic System for Transanal Microsurgery—Design and Validation *IEEE Robot. Autom. Lett.* **2** 1510–7

[14] Yung K L, Cheung J L K, Chung S W, Singh S and Yeung C K 2017 A Single-Port Robotic Platform for Laparoscopic Surgery with a Large Central Channel for Additional Instrument *Ann. Biomed. Eng.* **45** 2211–21

[15] Craig J J 2005 *Introduction to Robotics: Mechanics and Control* vol 3 (Pearson Prentice Hall Upper Saddle River)

[16] Mucksavage P, Kerbl D C, Pick D L, Lee J Y, McDougall E M and Louie M K 2011 Differences in grip forces among various robotic instruments and da Vinci surgical platforms *J. Endourol.* **25** 523–8

[17] Hwang M, Yang U-J, Kong D, Chung D G, Lim J, Lee D-H, Kim D H, Shin D, Jang T, Kim J W and Kwon D S 2017 A single port surgical robot system with novel elbow joint mechanism for high force transmission *Int. J. Med. Robot. Comput. Assist. Surg.* e1808

IOP Conf. Series: Materials Science and Engineering **320** (2018) 012009 doi:10.1088/1757-899X/320/1/012009

Adaptive Feedback in Local Coordinates for Real-time Vision-Based Motion Control Over Long Distances

M M Aref [1], P Astola[2], J Vihonen[3], I Tabus[2], R Ghabcheloo[1], and J Mattila[1]

[1] Laboratory of Automation and Hydraulics Engineering (AUT), Tampere University of Technology, FI-33101, Tampere, Finland
[2] Laboratory of Signal Processing (SGN), Tampere University of Technology, FI-33101, Tampere, Finland
[3] Cargotec Corporation, Tampere, Finland

Corresponding author: m.aref@ieee.org

Abstract. We studied the differences in noise-effects, depth-correlated behavior of sensors, and errors caused by mapping between coordinate systems in robotic applications of machine vision. In particular, the highly range-dependent noise densities for semi-unknown object detection were considered. An equation is proposed to adapt estimation rules to dramatic changes of noise over longer distances. This algorithm also benefits the smooth feedback of wheels to overcome variable latencies of visual perception feedback. Experimental evaluation of the integrated system is presented with/without the algorithm to highlight its effectiveness.

1. Introduction

In service robotics, the majority of tasks are based on robot intervention with known or unknown objects in indoor environments where global positioning systems are not accessible. Service robots benefit from using visual information to deal with objects their neighborhood. As reviewed in Walter *et al.* [1], state estimation and navigation for mobile platforms are commonly done using either global or local coordinates. Generally speaking, measurements using global coordinates (e.g. absolute position) are coarse and low frequency. Measurements using local coordinates, in which wheel odometry is available, are known to be smoother, fast and with a high resolution. In our case, the estimation from the environment is available in 30-40Hz while the dead-reckoning updates in 200Hz. To obtain smooth and globally valid feedback, fusion of these two types of measurements is a necessity but it is challenging and prone to drift [2], especially in heading estimation, which can affect coordinate mappings [3]. Some methods are based on stop, detect, look ahead and move [4]. However, our aim is to provide the object pose as a real-time feedback without the need for frequent replanning. Research that concerns itself with real-time detection [5] pays more attention to the precision of the image and less attention to the actual coordinates of the robot. For manipulation tasks, the repeatability of detection plays a vital role because the system might have physical interactions with the environment. Moreover, a path-follower controller with exact solutions needs a smooth and low-latency localization.
The object detection (ODT) results in this paper are comparable to Ha *et al.* [6]. However, in contrast with their conclusion, our method is capable of reaching smoother and more precise positioning of the robot with respect to the object without involving any image-based visual servoing in longer distances.

IOP Conf. Series: Materials Science and Engineering **320** (2018) 012009 doi:10.1088/1757-899X/320/1/012009

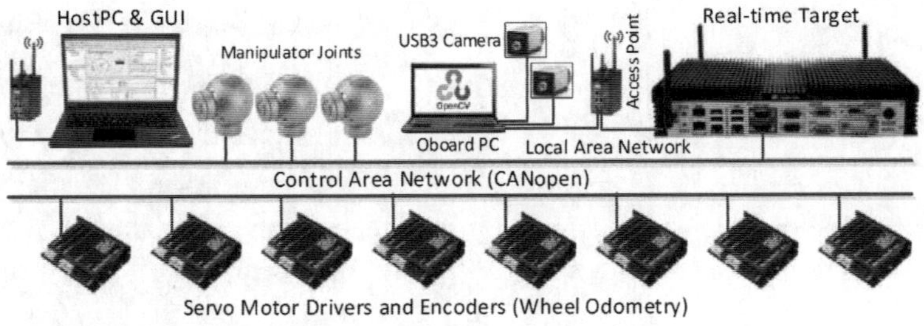

Figure 1. Control system's hardware architecture.

Figure 2. iMoro: The test case.

Long distance detections are important for non-holonomic mobile bases and autonomous vehicles because they need some space for steering and positioning before approaching the target area regardless of whether they are dealing with known markers or semi-unknown geometrical primitives.

1.1 Problem Definition and Architecture

The main goal of this research is to position a non-holonomic mobile manipulator in a location defined by target object(s). Each target object is recognizable by its longitude axis and geometric centroid, which builds a tensor of position/orientation ${}^{C}\mathbf{T}_{M}$ with respect to the camera frame. Since the result of ODT will be used as feedback for positioning tasks, it must be precise, smooth, well-behaved and available in real-time not only in terms of the image coordinates, but also in terms of the robot's coordinates and object's coordinates. This paper is a generalization of Aref *et al.* [7] for markerless detections.

Figure 1 illustrates the hardware architecture for the experiments and Figure 2 shows the test case robot with position gauge marker. Compared to Aref *et al.* [7], we use the ODT described in Sec.2.1 instead of ROS and ALVAR. This method benefits from the use of a stereo rig with two JAI-GO5000 USB3.0 cameras with a 0.16 meter baseline. Image grabbing and processing are done in C++ and OpenCV 2.4.9 developing environments. The sensor dimensions of the cameras are 2560x2048 pixels from which a cropped centered view of 1920x1080 pixels is used. Cameras operate at 35FPS, which was deemed as the shortest usable exposure time under the illumination conditions. These modules run on a Lenovo laptop with a Core i7 CPU. The latest results are transmitted to the *Real-Time Target* (RTT) on a LAN network. The algorithms described in Sec.2 are implemented on RTT.

2. Position Estimation for Manipulation in a Local Coordinate

The main aim of our project is to discover the potential of stereoscopic imaging for remote handling. The setup of stereo cameras was done according to the topology design, which takes view angle, stereoscopic precision, and lighting into account. Figure 3 describes the processing stages of the 3D Node module, i.e., the eye-in-hand vision system.

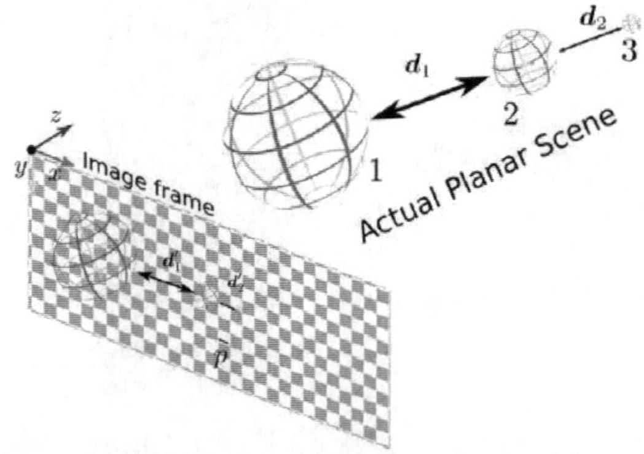

Figure 3. Estimation of distances on image frame can be comparable to pixel sizes if the objects appear in long distances.

Figure 4. Detection sample from left camera with object at 3.5m and 1m. The detected object contour is illustrated partially with a white line.

2.1 Object Detection (ODT)

ODT follows that of Astola *et al.* [8] where a stereo camera system was used for object detection in 30-40Hz and pose estimation. In their research, object detection was based on color image thresholding in the hue channel and subsequent contour estimations. Object pose estimation performs stereo matching on the contour to derive the pose of the object in 3D coordinates. In Astola *et al.* [8], two versions of contour processing were studied – one using the convex hull and the other using the full contour. While the convex hull version performed faster computationally, the use of full contours with an additional filtering step provided superior results.

The output of ODT is the object pose tensor ${}^{C}\mathbf{T}_{M}$, which is constructed by the pose estimation algorithm using the centroid and orientation of the detected object. In Astola *et al.* [8], the immediate goal of the ODT was to maintain high enough rate of detection and pose estimation, as the sensor fusion prefers high frequency outputs from ODT. As a consequence, virtually no filtering or image processing is done jointly on the acquired frames and pose estimates. This results in a fair amount of high frequency pose estimation noise originating from ODT. The accuracy of the pose estimation is heavily range dependent, as the object's dimensions in the image plane diminish as a function of distance. Additionally, for a fixed stereo baseline, the system's ability to accurately estimate object depth also decreases with distance. The distance-based non-linearities in pose estimation, especially in the heading, are accommodated in a novel sensor fusion approach.

2.2 Experimental Results on Open-loop

To compare open-loop performance we used a controller independent of the vision feedback, which smoothly drives the robot in a straight line. It accelerates in the beginning and then maintains a constant velocity. As shown in Figure 4, if both systems are given the same input data, the M-Estimator is capable of generating smoother feedback. The acceleration moment of the robot is magnified on the graph to highlight the fast and smooth reaction of the filter. As illustrated, while the M-Estimator has faster motion feedback, it remains smoother across longer distances.

2.3 Integration in feedback

Exploiting the results of ODT in a closed-loop control system requires smoothness, availability and consistency of data at each sample time. Initially, an ODT module does not have such characteristics. Therefore, in this section we propose an algorithm that can integrate the ODT result in a real-time controller loop. To highlight the differences, we compare the following subsystems at each step:

- Variation of Object Detection Algorithms: Ver.1 and 2 in Sec.2.1, Ver.2 in Sec.2.1, a marker-based AR tag detection method [3].
- Modifications of M-Estimator: Gain adaptation to measurement changes in Sec.2.4, and gain adaptation to depth in Sec.2.5.

2.4 M-Estimation for Real-time Feedback

For each component of ${}^{C}\boldsymbol{T}_{M}$, such as y, while we are mapping all the vectors into the object's local coordinate $\{W\}$, we also apply the following estimation equations to estimate z:

$$\frac{\partial}{\partial t}\begin{bmatrix}\hat{z}\\ \hat{b}\end{bmatrix} = \begin{bmatrix}0 & 1\\ 0 & 0\end{bmatrix}\begin{bmatrix}\hat{z}\\ \hat{b}\end{bmatrix} + w\begin{bmatrix}k_P\\ k_I\end{bmatrix}(y - \hat{z}) + \begin{bmatrix}1\\ 0\end{bmatrix}x \tag{1}$$

$$w = \exp\left(-\left(\frac{\|\hat{y} - z\|}{c}\right)^2\right) \tag{2}$$

where x is the corresponding velocity component from the interoceptive sensor, such as wheel odometery [7]. Moreover, w accommodates the varying nose statistics by introducing the so-called robust Welsch function. Although this filter is capable of generating a smooth and robust feedback for marker-based ODT, for marker-less ODT the parasitic disturbances are heavily range-dependent (as shown in Figures 4 and 5) and should be accommodated accordingly.

2.5 Adaptive Filtering

As shown in Figures 4 and 5, the method described in 2.4 is capable of rejecting noises and disturbances initially. However, across longer distances, the object's image size becomes too small (e.g. Figure 4) and detection characteristics are different, especially for the orientations. For the scene illustrated in Figure 3, assume that triangulation between the image frame and actual object places in a pin-hole camera model is supposed to be valid for objects 1, 2 and 3. Then, the ratio of d'_1/z_1 linearly affects the tangent of the object orientation. Practically, this deterministic calculation is not valid for object places such as 3, where the distance d_2 is comparable to real pixel wideness, p. Quantization errors simply can be equal to or bigger than p and signal to noise ratio is considerable. This

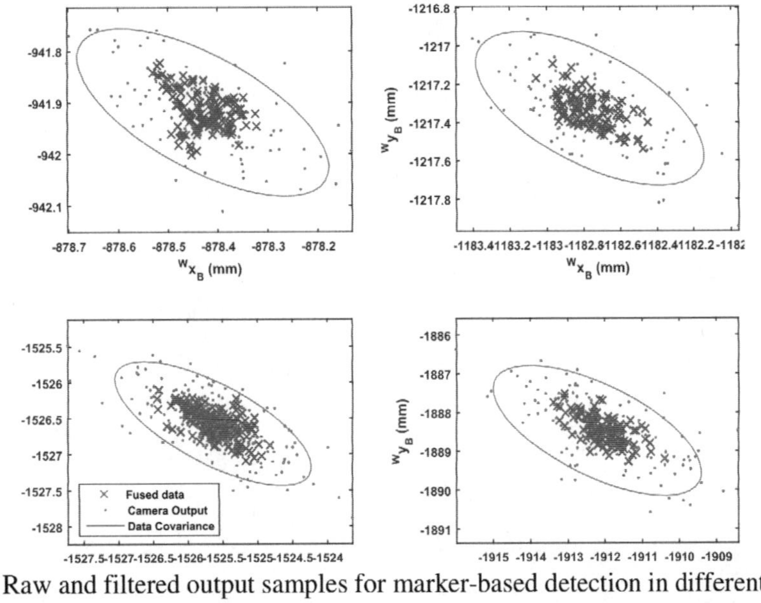

Figure 5. Raw and filtered output samples for marker-based detection in different distances. The numbers represent estimated position of the robot in target's coordinate frame.

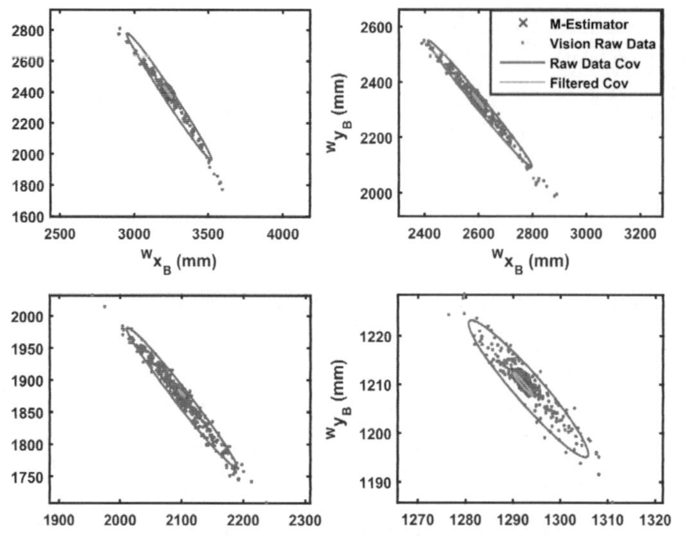

Figure 6. Results of similar tests to Figure 4 but for marker-less detection. Centroids of target objects rely on the origin of the coordinate system.

condition becomes worse if object 3 is placed too far from the adjusted focus range of the camera's optical system. This is the place where the pin-hole model is not necessarily correct, and d'_1/d_1 and d'_2/d_2 ratios do not exactly obey triangulation rules. Therefore, we consider correlation between the depth and accuracy of all estimated values.
As shown in Figure 5, noise amplitude in the first quarter of the path is comparable to the outliers of the last quarter. These changes result in larger values of $\|\hat{y}-z\|$ in (2). Therefore, w gain of (1), which is supposed to prevent integration of unexpected measurements in longer distances, prevents convergence to the final value, as shown in Figure 6. and the left graph of Figure 7. From the lessons we learned during the experimental studies on the system, the Welsch weight should be adaptive to the camera-object distance,

$$c = k_{w1}\exp\left(k_{w2}(\hat{d}-d_{min})\right)+k_{w3} \tag{3}$$

where $\hat{d}$ is the estimated distance to the object and d_{min} is the minimum detection distance where we want value c to converge to $k_{w1}+k_{w3}$. Usually, this distance is the grasping point. $k_{w1}k_{w2}$ adjusts the rate of sensitivity of w to the amplitude of noises.

According to the experiments, for Ver.1, there is also a significant change of the mean values while approaching the object. Thus, we can conclude that Ver.1 is not useful for the closed-loop control loop as it adds significant non-linear behavior to the position estimation in local coordinates. So, (1) is suitable for the ODTs similar to Ver.2 where the quantization errors do have a dramatic effect on the convergence rate. For these ODTs, substitution of (1) in (2) compensates the dependency of noise dynamics on the distance. As shown in Figure 7b, the adaptive M-Estimator is capable of following the camera output even after a considerable amount of outliers. These methods provide the robot with real-time feedback of position suitable for closed-loop position-based visual servoing as shown in Figure 8 by adaptation of estimation weights. Some of the tests are also demonstrated on the YouTube video[1].

---

[1] Object tracking on the go, by iMoro https://www.youtube.com/watch?v=m4vsr1rvxUg

3. Conclusions

In this paper, by combining our previous experience on machine vision application in non-holonomic steerable platforms for mobile manipulation, we modified the widely-applied principles of machine vision, and complementary and Kalman filtering, for pose estimation. On one hand, we demonstrated how modifications in ODT can reduce non-linearity of detection for unknown objects. On the other hand, by proposing an adaptive estimation method, we achieved nearly the same quality of mobile robot positioning compared to marker-based ODT in distances that had been considered too far for position-based motion control. It is important to note that long range ODT and position estimation is essential for non-holonomic robots to keep their maneuvers smooth without stopping and replanning. The results are experimentally demonstrated on a video for detection, estimation and positioning on the go. Extension of this work for higher degrees of freedom is a future work.

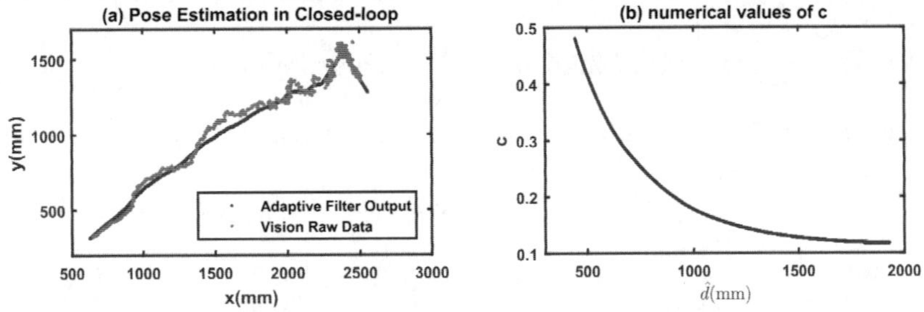

(a) Pose estimation in the object's coordinate (b) Changes in the c value as

Figure 7. Path and fusion behavior is adapted by distance from the target because values are functions of range in (3) and the robot adjusts the weighting in (2) while approaching to the target.

4. References

[1] Walter M R *et al.* 2015 A situationally aware voice-commandable robotic forklift working alongside people in unstructured outdoor environments *J. of Field Robotics* 32 pp 590–628

[2] Teller S *et al.* 2010 A voice-commandable robotic forklift working alongside humans in minimally-prepared outdoor environments *IEEE Int. Conf. on Robotics and Automation* pp 526–33

[3] Wang Y and Wang S 2017 A new directional-intent recognition method for walking training using an omnidirectional robot *J. of Intelligent & Robotic Systems* pp 1–16

[4] Wu J, Yue X and Li W 2016 Integration of hardware and software designs for object grasping and transportation by a mobile robot with navigation guidance via a unique bearing-alignment mechanism *IEEE/ASME Trans. on Mechatronics* 21 pp 576–83

[5] Jung B and Sukhatme G S 2010 Real-time motion tracking from a mobile robot *Int. J. of Social Robotics* 2 pp 63–78

[6] Ha H U, Yoon H N, Kim Y K, Lee D H and Lee J M 2015 Object tracking of the mobile robot using the stereo camera *Int. Conf. on Intelligent Robotics and Applications* pp 322–330

[7] Aref M M, Vihonen J, Ghabcheloo R and Mattila J 2018 On latencies and noise effects in vision-based control of mobile robots *Advances in Service and Industrial Robotics: Mechanism and Machine Science* 49 pp 191–9

[8] Astola P, Aref M M, Vihonen J, Mattila J and Tabus I 2017 Object detection in robotic applications for real-time localization using semi-unknown objects *40th Int. Conf. on Telecommunications and Signal Processing*

5. Acknowledgements

This work is supported by the Academy of Finland under the grant numbers 286260 and 283171.

Design of Precise Control and Dynamic Simulation of Manipulator for Die-casting Mould

Yi Mei[1], Chuang Liu[2], Chunlan Zhu[3], Quanlong Sun [4]

[1]Professor, Department of Material Engineering, School of Mechanical Engineering , Guizhou University, China.
[2,3,4] Master degree, School of Mechanical Engineering, Guizhou University, China.

Abstract. To meet the demand of the enterprise for the die-casting machine, he overall structure of the manipulator and the controller of each joint were designed. Firstly, The solid model of the manipulator is established; and then, the simplified manipulator model was for dynamics simulation; Finally, through a combination of structural dynamics and dynamics control, established the cooperative control system of the manipulator, designed parameters of each joint motion controller and applied to cooperative control system simulation, joint simulation of manipulator is realized .The simulation results show that the manipulator system has good fast response characteristics and trajectory tracking characteristics, and the overall force results are in accordance with the theory.

1. Introduction

During the die-casting production, in order to save costs and achieve automated production, enterprises needs of the robot are growing. As we know robot is a complex system, the traditional mechanical design means in the structural design and control system design can not meet the requirements, so it would be necessary to build a virtual prototyping system for a robot ,before creating a solid prototype[1-2]. Wu Xing [3] has used multi-objective genetic algorithm to optimize the servo controller PID parameter method to achieve servo. Yu Xiao[4] has used POS optimization algorithm to achieve the optimization of the robot parameters. Chen Gang[5] has studied the joint simulation of 7-DOF manipulator based on ADAMS and MATLAB, and the dynamic response characteristics of the manipulator are obtained by the virtual prototype system analysis. Chen Shuijin[6] used the method of co simulation of virtual prototype based on ADAMS and MATLAB to establish a three degree of freedom manipulator model. The joint simulation control of virtual prototyping is helpful to shorten the overall cycle of the design and improve the overall performance of the manipulator, which is beneficial to the overall research and development of the manipulator.

2. Structural design and dynamic analysis of manipulator

2.1. Structural Design and Working Principle of Manipulator

According to the actual environment of enterprise production and demand conditions, the simplified model of the manipulator design process including a total of five pairs, of which four rotations and a ball screw (in the simulation will be simplified for the mobile side between the rail and the slider), and to establish three-dimensional model in UG environment, such as shown in figure 1:

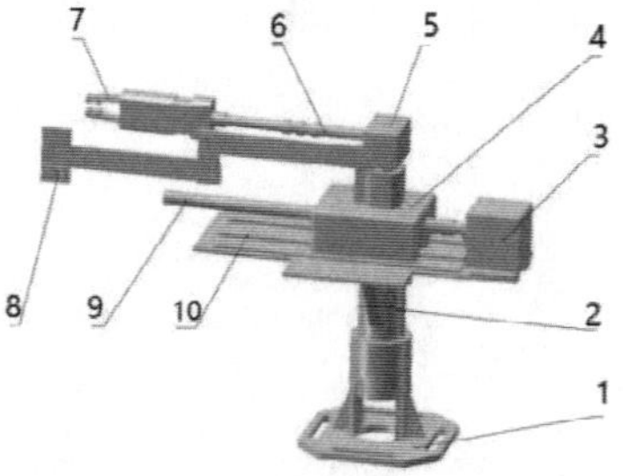

1 Machine base 2 Support frame 3 Tail motor 4 Slider 5 Rotating cap
6 Mechanical arm 7 Mechanical gripper 8 nozzle 9 Screw 10 guide rail
Figure 1.manipulator structure.

The main design parameters of the manipulator are shown in Table 1.

Table 1. Main parameters of manipulator .

parameters	Values
height(mm)	1544
length（mm）	1740
width(mm)	765
arm length(mm)	1008
guide rail length(mm)	1250
working space(mm^3)	1405×1008×60
Working period(s)	15
Work efficiency	2/15s

The main working principle of the manipulator: The support frame 2 can be moved up and down or rotation according to the actual production , and it can not be considered once the position has been determined.

The working principle of the manipulator can be simplified as follows: the tail motor 3 drives the screw 9 to rotate, and the slider system 4 is moved back and forth on the guide rail 10 by the ball screw .After the mechanical gripper 7 getting the mould parts , the nozzle 8 is moved back and forth the slider sys-tem to clean the mould parts, and then rotating cap 5 is rotated with the mechanical arm 6 and the mechanical gripper 7 puts down the parts .Finally ,the entire system back to the initial position, to achieve periodic pickup movement.

2.2 Dynamics Simulation of the Manipulator

By defining the material of each component, the moment of inertia and other related properties, the motion pairs are set up and the relevant drivers are used to ensure that the model import process is not distorted and the mechanical movement is effective. And according to the design trajectory of the manipulator, the driving function of the manipulator is set up. Among them, the mechanical gripper's motion function is taken as an example: STEP(time, 0 , 0 , 1 , -0.23) +STEP(time , 2.5 , 0 , 3 , 0.23)+STEP(time , 8.5 , 0 , 9 , -0.23) ,The simulation results , including the angular velocity, velocity and displacement curve of motion simulation, and the trajectory of the mechanical gripper are in good agreement with the design expectations, which indicates that the connection and driver settings are reasonable. The manipulator model can simulate the motion according to the given driving function. The result shows that it can be used as the subsystem of the next joint simulation.

3. The design of the dynamic system

To achieve the coordination simulation of the control system and the dynamics system, it is necessary to introduce the dynamics model of the manipulator into Matlab, and the dynamics system works as the mechanical sub-model of the control system so that the data can be successfully exchanged . In particular, 20 state variables are established through the Adams / control module, including four torque variables, a driving force variable, 10 position variables, 5 (angle) speed variable. The specific settings are shown in Table 2

Table 2. Settings of the State variable

Num	NAME	Attributes
1	DGripperRotateOme	INPUT
2	DGripperRotateAng	INPUT
3	UpGripperRotateOme	INPUT
4	UpGripperRotateAng	INPUT
5	Rotating cap RotateOme	INPUT
6	Rotating cap RotateAng	INPUT
7	Mechanical arm RotateOme	INPUT
8	Mechanical arm RotateAng	INPUT
9	Slider MoveVel	INPUT
10	Slider MovePos	INPUT
11	Torque_DGripper	OUTPUT
12	Torque_UpGripper	OUTPUT
13	Torque_Mechanical arm	OUTPUT
14	Torque_Rotating cap	OUTPUT
15	Force_Slider	OUTPUT
16	DGripperAng	OUTPUT
17	UpGripperAng	OUTPUT
18	Rotating capRotateAng	OUTPUT
19	Mechanical armRotateAng	OUTPUT
20	Slider MovePos	OUTPUT

The motion of the joint is driven according to the instruction of the variable, and then a complete closed loop control is realized through the feedback system to realize the precise control. After the status variable setting is complete, 10 input variables and 10 output variables are set up and showed in the control system, The mechanical subsystem shown in Figure 2 and Figure 3

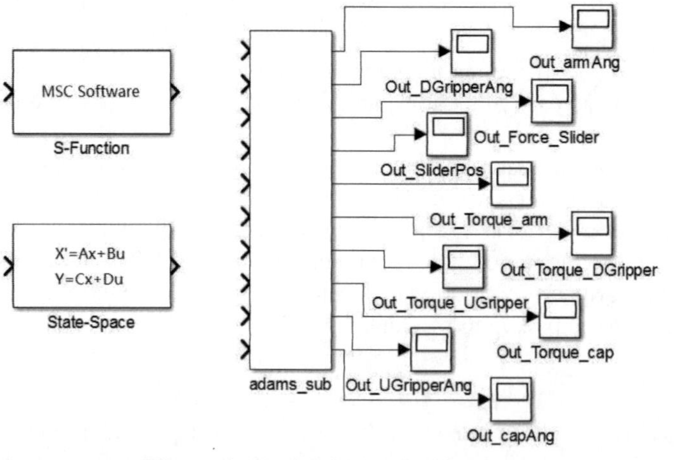

Figure 2. Mechanical sub system.

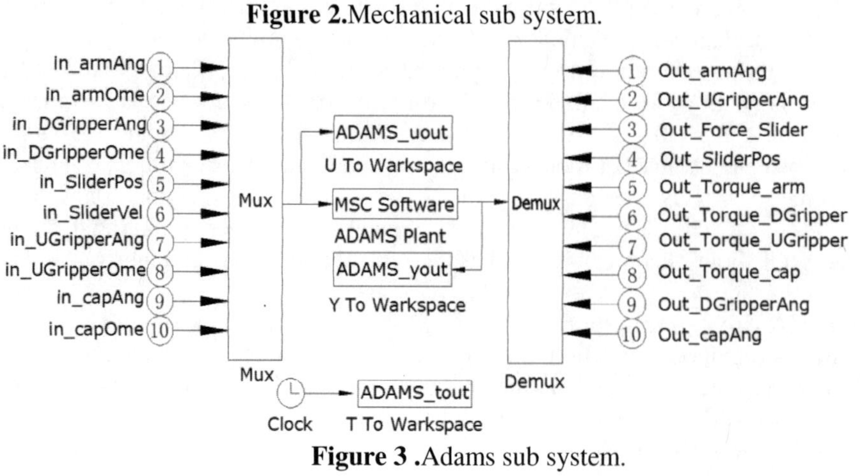

Figure 3 . Adams sub system.

IOP Conf. Series: Materials Science and Engineering **320** (2018) 012010 doi:10.1088/1757-899X/320/1/012010

4. The design of the control system

The manipulator system has the characteristics of strong coupling, which should be passed into a linear multivariable system in the process of joint simulation. The PD controller is used to control the movement of each joint to realize the precise control of the whole manipulator[7-8].

Specifically, the manipulator is controlled by PD in ADAMS, and the actual motion is planned in SIMULINK, the control principle is as follows:

$$\mathrm{F} = k_p(x_d - x_m) + k_d(\dot{x_d} - \dot{x_m}) \quad (1)$$

$$\tau = k_p(\theta_d - \theta_m) + k_d(\dot{\theta_d} - \dot{\theta_m}) \quad (2)$$

Among them, x_d, $\dot{x_d}$, θ_d and $\dot{\theta_d}$ are locations and speeds of planning in simulink, x_m, $\dot{x_m}$, θ_m and $\dot{\theta_m}$ are the actual measured values in ADAMS. Due to the influence of gravity on the mechanical gripper, the controller requires gravity compensation, which has little effect on the overall controller of the manipulator and is complicated. In order to reduce the interference of the gravity to the analysis results, the quality of the mechanical grippes are reduced to 10% of its true quality. The control system of the joint simulation as follows:

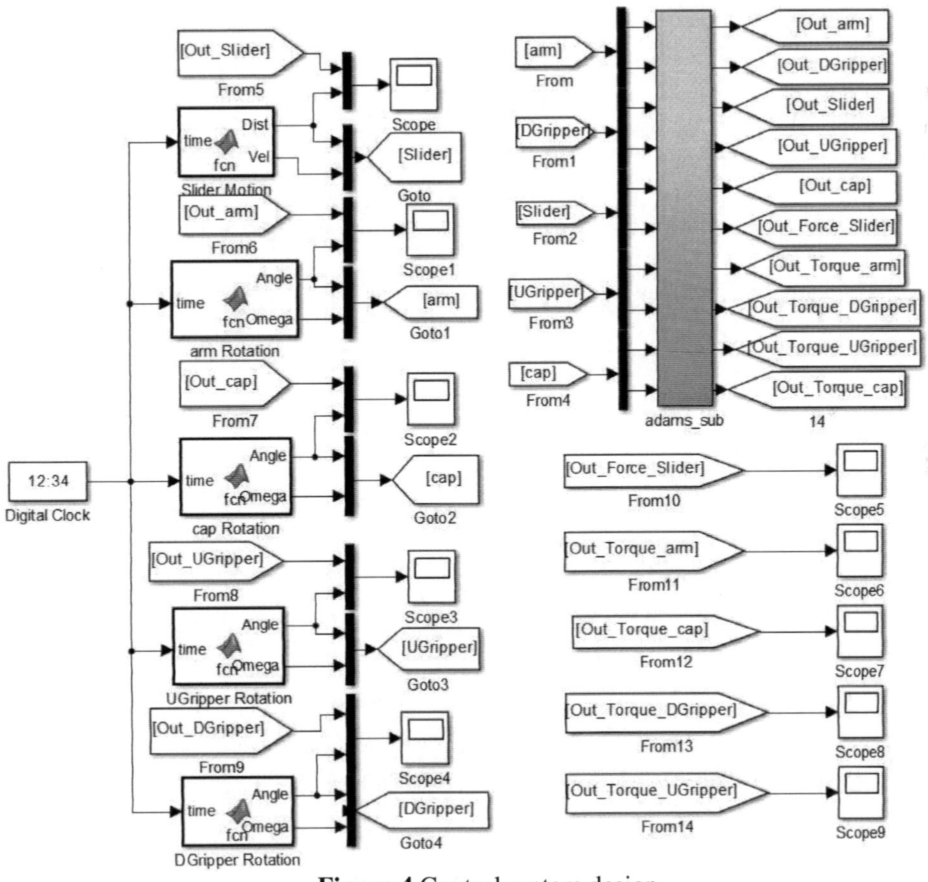

Figure 4. Control system design.

The Joint simulation control system using PD controller is divided into three sections, including the ADAMS dynamics module, Simulink function frame and related results. Firstly, the motion functions of each joint movement are designed though the FCN module of SIMULINK, and then the functions are input to ADAMS dynamics module. Finally, the trajectory and force condition of the main joint of the manipulator are obtained by simulation

5. The Realization and Result Analysis of Joint Simulation

In order to verify the accuracy and rationality of the design of the robot joint simulation control system, run Simulink and set the step length to 0.005, and set the simulation time to 15 s. FIG. 5 FIG. 8 ows the input (Angle) displacement curve of the main moving parts of the manipulator and the tracking (angular) displacement curve after its simulation. (the movement of the lower jaw is essentially the same as that of the upper claw, just study the movement of the upper paw in this case)

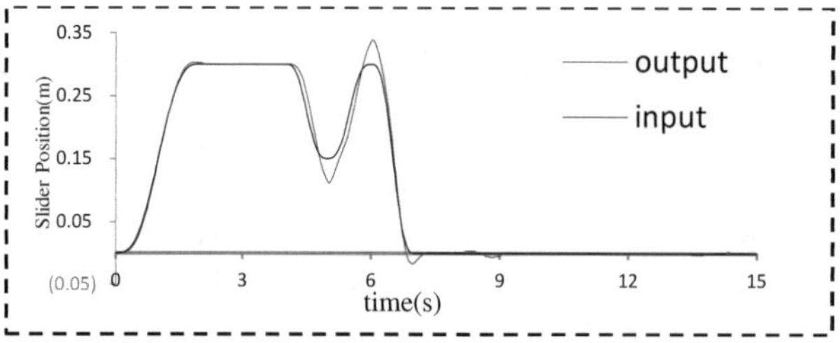

Figure 5. The displacement of the Slider.

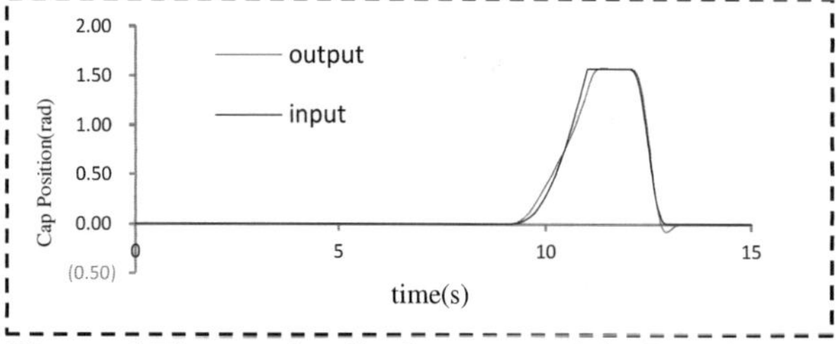

Figure 6. The angular displacement of the Rotating cap.

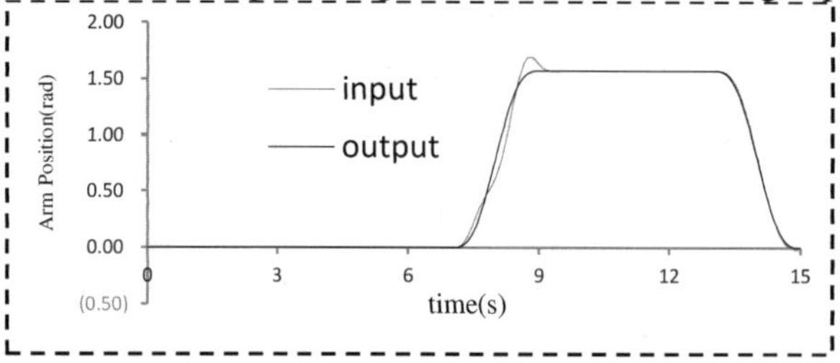

Figure 7. The angular displacement of the Mechanical arm.

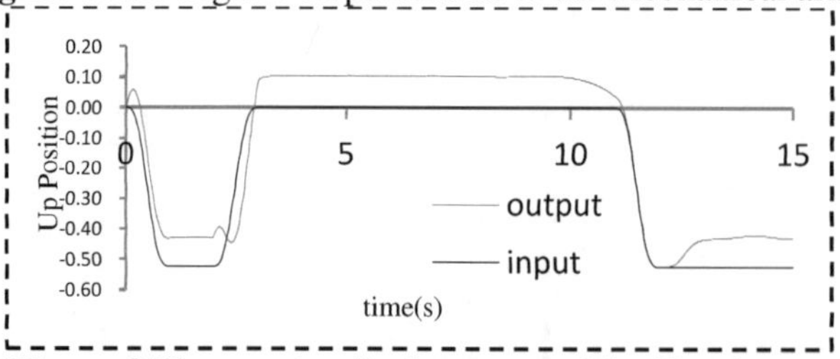

Figure 8. The angular displacement of the Upgripper.

Compare and analyze The expected trajectory of the design and The trajectory of the simulation results of the main components, and the overshoot was not more than 16.7%.

It shows that the design of the controller can meet the requirements of the controller's fast response to the controller and the precise trajectory tracking.

The input variable of the controller is the (angular) displacement and (angular) velocity of each joint. Through the joint simulation, the force of each joint in the motion is obtained. FIG. 9 - FIG12 shows the force and withstand torque of each joint component.

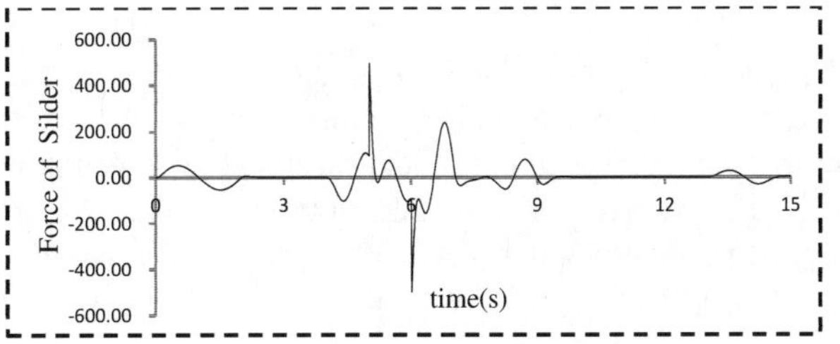

Figure 9. The force of the Slider.

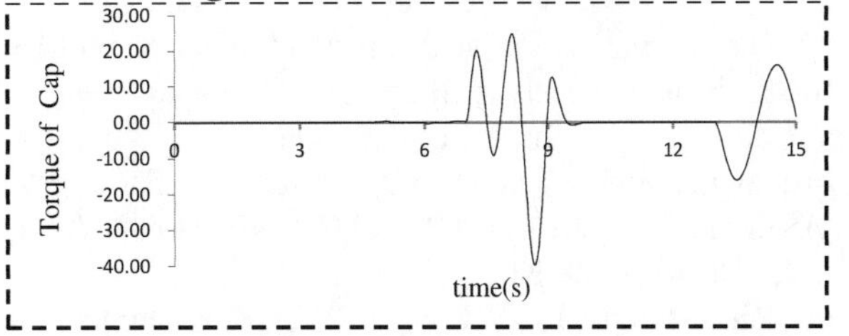

Figure 10. The torque of the Rotating cap.

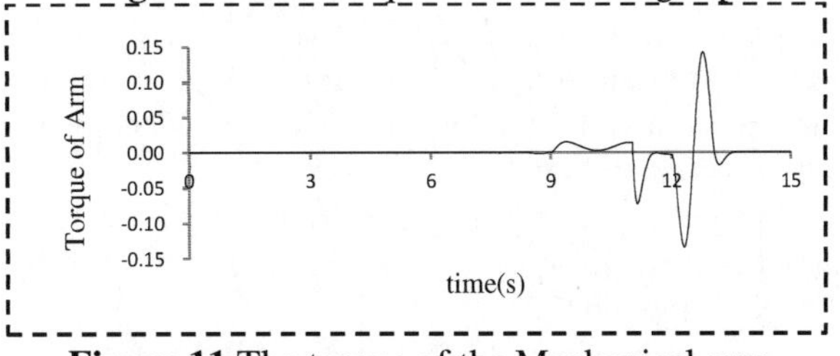

Figure 11. The torque of the Mechanical arm.

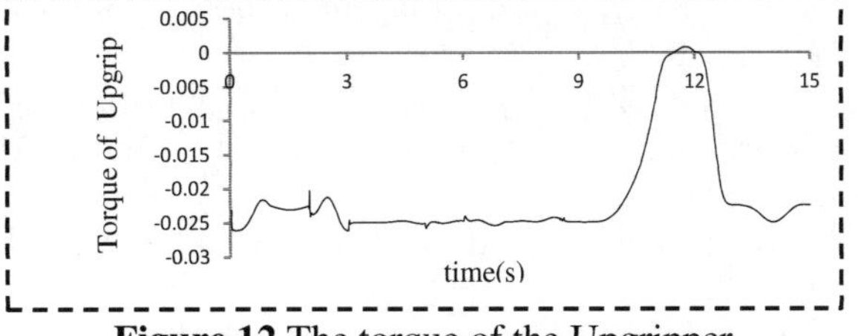

Figure 12. The torque of the Upgripper.

(Angle) Displacement curve of the main part is combined with the force (torque) case to analysis. Taking the slider as an example, the displacement function of the slider as the following:

0-2s

$$\text{Dist}=(0.3/2)*(\text{time}-2)/(2*\text{pi})*\sin(2*\text{pi}*\text{time}/2)) \tag{3}$$

2-4s

$$\text{Dist}=0.3 \tag{4}$$

4-5s

$$\text{Dist}=0.3-(0.3/2)*(\text{time}-4)-1/(2*\text{pi})*\sin(2*\text{pi}*(\text{time}-4)))) \tag{5}$$

5-6s

$$\text{Dist}=(0.3/2)+(0.3/2*(\text{time}-5)-1/(2*\text{pi})*\sin(2*\text{pi}*(\text{time}-5)))) \tag{6}$$

6-7s

$$\text{Dist}=0.3-(0.3/2*(\text{time}-6)-1/(2*\text{pi})*\sin(2*\text{pi}*(\text{time}-6))) \tag{7}$$

Compare and analyze the displacement driven functions of the slider input and The forces curve of the slider , basically in line with the theory of the force of the slider.

Therefore, a joint simulation system based on Adams and matlab can be used to simulate the movement of manipulator preferably.

6. Summary

The motion simulation model uses the PD controllers, and the results of joint simulation analysis show that the controller has good fast response and trajectory tracking characteristics, and finally gets the force of the main joints, which lays the foundation for the development of manipulator entity prototype. On this basis, in order to improve the motion accuracy of the manipulator, reduce the vibration of the manipulator, the focus of future research is the movement of each joint motion optimization parameters, to improve the motion accuracy of manipulator in the engineering work.

7. References

[1] Zhao Yan 2014 The simulation of Quadruped Robot Based on MATLAB and ADAMS .Master Thesis of ShanDong University.

[2] Zhao Daxing , Yang Yong1 , Xu Wan , Ding Guolong and Peng Ling 2014, Mechanical and Electrical Co-simulation for CNC Machine Tool Feed Drive System Based on ADAMS and MATLAB MACHINE TOOL & HYDRAULICS (May 2014 Vol. 42 No. 10)pp 28-31

[3] Wu Xing , Lou Peihuang and Tang Dunbing 2011 Integrated Motion Control of Path Tracking and Servo Control for an Automated Guided Vehicle JOURNAL OF MECHANICAL ENGINEERING(Feb 2013 Vol. 47 No. 3)pp 43-48

[4] Yu Xiao, Liu Dong and Lan Weiyao 2011 Parameters Optimion of Robot Arm PID Controllers and on POS Algorithm Modular Machine tTools & Automatic Machining Technology 02(2011)pp 89-92

[5] Chen Gang, Zhou Qicai, Wu Jing and Yan Nan 2017 Research for Co-simulation of 7DOF Manipulator Based on Adams and Matlab Journal of System Simulation (Jan 2017 Vol29 No. 1)pp 99-106

[6] Chen Shuijing ,and Wu Yuxiang 2012 Adaptive Neural Control of Robot Manipulators and Virtual Realizing Modular Machine tTools & Automatic Machining Technology 03(2012)pp 50-56

[7] Ma Ruqi, Hao Shuanghui, Zheng Weifeng, Hao Minghui and Song Baoyu 2010 Research on coordinated simulation of robot arm based on MATLAB and ADAMS Machinery Design & Manufacture 04(2010)pp 93-95.

[8] Li Yueyue. Robet simulation based on ADAMS and MATLAB. Master Thesis of Hebei University.2010

IOP Conf. Series: Materials Science and Engineering **320** (2018) 012011 doi:10.1088/1757-899X/320/1/012011

Multi-function robots with speech interaction and emotion feedback

Hongyu Wang, Guanting Lou and Mengchao Ma

School of Instrument Science and Opto-electronic Engineering, Hefei University of Technology, Hefei, Anhui 230009, P.R. China

Corresponding author's e-mail: louguanting@gmail.com

Abstract. Nowadays, the service robots have been applied in many public circumstances; however, most of them still don't have the function of speech interaction, especially the function of speech-emotion interaction feedback. To make the robot more humanoid, Arduino microcontroller was used in this study for the speech recognition module and servo motor control module to achieve the functions of the robot's speech interaction and emotion feedback. In addition, W5100 was adopted for network connection to achieve information transmission via Internet, providing broad application prospects for the robot in the area of Internet of Things (IoT).

1. Introduction

With the rapid development of China in recent years, robots in the manufacturing and industrial areas have been applied in large quantities, but their applications in the service industry are not particularly extensive yet. In recent years, there have been cleaning robots and cooking robots emerging in the market, indicating that it has been realized that service-oriented robots will be a very important application in the future. However, most of the robots, which has been put into use in the current domestic services, can only complete a single function and they do not have the ability to communicate with people and cannot provide emotional feedback either. At the international market, the research on service-oriented robots in Japan has reached a very high level. The "HRP-2" developed by the Japanese Industrial Technology Research Institute and the Shimizu Construction Company and the "Small IF" developed by the Japanese Human Robot Foundation have not only the ability to communicate with humans, but also the ability to speak, they all represent the current highest level in the development of service-oriented robots [1].

Based on these realities, under the premise of widely used Internet technology, this investigation designed and implemented a multi-function robot with speech interaction and emotion feedback. Different from the current majority of service-oriented robot, the robot can do speech recognition and face expression change. Those would not only meet the user's emotional and operational requirements, but also set the dialogue content and feedback mechanism in a specific occasion and realize some of the specialized services in public places. It is another prominent feature of its internal work instructions are achieved by the open source development platform. Therefore, customers can set their own parameters according to their needs and achieve its function and application at a great convenience. The diversity of the scene also allows users to personally participate in their own robot partners "recycling" process.

2. Realization of speech interaction function

Automatic speech recognition (ASR) is a technology that transforms a speech signal into a corresponding text file or command by identifying and understanding the process [2]. Speech recognition is the key technology for speech interaction robots, while the dialogue between human being and robot in natural language are realized. In order to achieve the speech interaction function, ASR M08-A non-specific speech recognition module was used in this study.

2.1. Work principle for ASR M08-A module

LD3320 speech recognition chip functions as a core of ASR M08-A non-specific speech recognition module. The chip is designed and manufactured by ICRoute company, integrates speech recognition processor and some external circuits including AD, DA converter, microphone interface, sound output interface, and so on. The all set-up do not have any external auxiliary chip, whereas speech recognition/speech/man-machine dialogue functions can be realized [3-4]. Also, the identified list of key words can be edited dynamically, which provides a basis for its wide application in the field of robots.

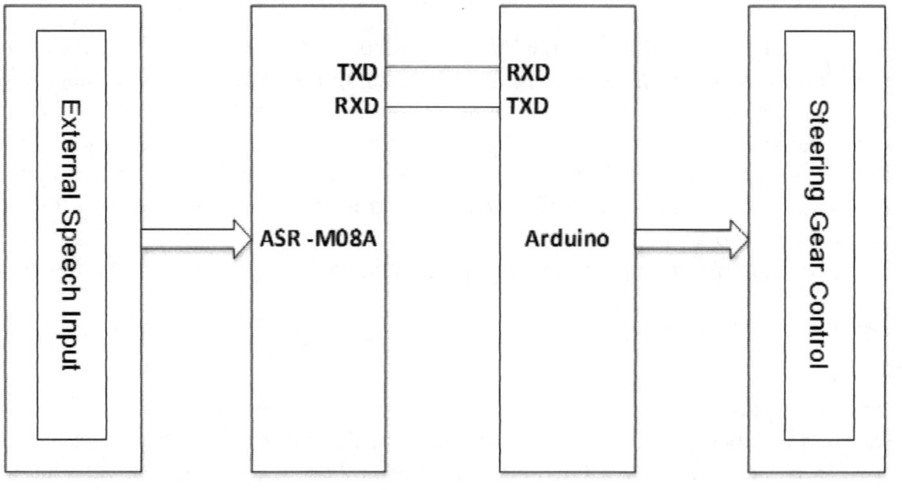

Figure 1. The connection details between speech modules

2.2. Hardware design of speech recognition function

ASR M08-A non-specific speech recognition module has total of six external serial port, namely SPK-, SPK +, 3.3V, TXD, RXD and GND. Wherein SPK is the interface of the external speakers, 3.3V and GND are the positive and negative terminals of the module power supply, TXD and RXD are the communication interface of the modules. In normal operation, the module receives the external voice signal input, through the LD3320 chip into a serial output signal, and sent to the Arduino microcontroller. The specific implementation steps are shown in Figure 1.

3. Realization of emotional feedback function

In order to achieve the emotional feedback function, a steering gear is connected to the robot's eyebrows, eyes, mouth, eyelids as shown in Figure 2. With the coordination of the steering control, the speech interaction is achieved and at the same time this unique configuration allows the robot to demonstrate emotion feedback as well.

Figure 2. Face diagram of the multi-function robot

3.1. Steering gear control

The control signal of the servo is a pulse width modulation (PWM) signal with a period of 20ms, where the pulse width is in the range of 0.5-2.5ms and the corresponding rudder position with the freedom of linear change can be 0-180 degrees. In other words, at a giving pulse width, its output shaft will remain at the corresponding angle, no matter how the external torque changes. The rudder will change the output angle to a new correspondence position until another pulse is provided,

The control of the steering gear generally requires a time base pulse of about 20ms, the time base pulse is in the range of 0.5ms-2.5ms. The corresponding relationship of the steering gear and time base pulse width is shown in Table 1:

Table 1. The correlation of the time pulse width and the steering gear position

Time Base Pulse Width (ms)	Steering Gear Position (degree)
0.5	0
1.0	45
1.5	90
2.0	135
2.5	180

The steering gear has a reference circuit inside to produce a reference signal with a period of 20ms and a width of 1.5ms. There is a comparator that compares the applied signal with the reference signal to determine the direction and size to produce the motor's rotation signal. It can be seen that the steering gear is a position servo drive, the rotation range cannot exceed 180 degrees. The steering gear is the key device to realize the change of expression and maintain multi-functions for the investigated robot.

Second, in order to be able to coordinate the control of multiple steering gears, a 32-way servo controller is used to achieve the corresponding function. The 32-way servo controller connected to the steering gear control chip functions as the core control unit. Therefore, 32 external steering gears can be integrated for simultaneous control with conveniently adding the program according to the requirements in the corresponding control software. Meanwhile, the user can customize the instruction based on such configuration. The 32-way servo controller provides a variety of instructions, which can be distributed into the following three categories as shown in Table 2:

Table 2. Commands of the steering gear controller

Component Name	Commands
Single Steering Gear	#1P1500T1000\r\n
Multiple Steering Gear	#1P1500#2P1500T1000\r\n
Action Chains	G1F3\r\r

For a single commander and a control command for multiple servos, the numbers 1 and 2 after "#" are the number of the steering gears. The number 1500 after "P" is the steering angle value in the range of

IOP Conf. Series: Materials Science and Engineering **320** (2018) 012011 doi:10.1088/1757-899X/320/1/012011

500-2500. The number 1000 after "T" is the speed of the steering gear in the range of 100-9999. For action group instructions, the number 1 after "G" represents the action group number, and the number 3 after "F" represents the cycle of executions of the action group.
In this study, the action group instructions are applied the steering gear number to adjust the position of the steering gear, and then add the action group. The dynamic process is achieved by the cycle number of operations. The specific program codes see the section 4.2 control system software design in this article.

3.2. Hardware design of emotional feedback function

Arduino receives the data from the ASR M08-A module and converts them to the 32-way steering controller through the serial port. The TXD and RXD serial ports on the 32-way servo controller functions as the serial port for data exchange with the Arduino microcontroller, and control the steering gear for the corresponding action.

4. Application of network module

W5100 network expansion module [5] directly uses the Ethernet library file in the Arduino IDE development environment and makes it as a simple Web server. Read and writing the digital and analog serial port and other network applications can be achieved through the network control as well.
In this study, the W5100 module is used as a simple self-service hotel service system. The robot will automatically help to realize the corresponding function by saying the corresponding instructions such as "room reservation " and "check room" as well as the corresponding room on the webpage is listed as "subscribed", "can be booked" and so on (web page program to rely on JavaScript language to complete).

5. Design of control system

The core of the control system used in this study is the Arduino microcontroller. Arduino is an open source electronic design platform. The hardware part is consisted of an Atmel AVR microcontroller, I / O interface and related circuits. The software part includes the standard program compiler and program downloader, with similar Java and C language Processing / Wiring Development environment [6-7]. The main controller used in this study is Arduino Mega2560. Arduino Mega2560 core is ATmega2560 microcontroller, which has 54 digital input / output interface (16 channels can be used as PWM output), 16 analog input, 4 UART interface, a 16 MHz crystal oscillator, a USB port, and a power outlet, an ICSP header and a reset button. The control board is connected to a computer via standard USB cable for easy program debugging [8].

5.1. Design of control system hardware

Arduino Mega2560 as the core of the control system, including the ASR M08-A non-specific speech recognition module, 32-way servo controller, a number of steering gear, and W5100 network module. The circuit wiring diagram is shown in Figure 3.

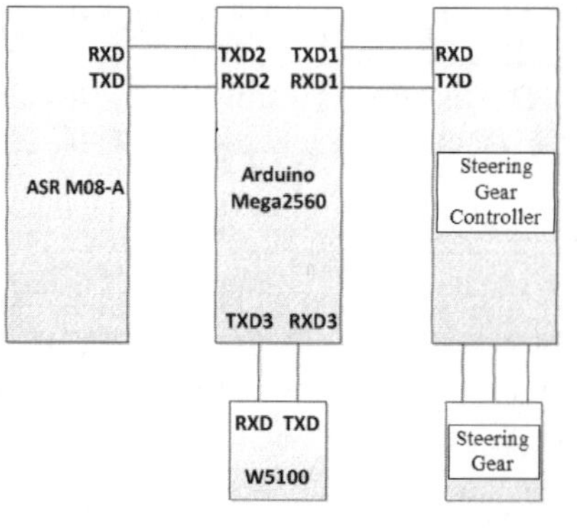

Figure 3. The connection details of hardware components

5.2. Design of control system software

This study makes use of the Arduino IDE development environment combined with the JavaScript language written by the web program to achieve the simple application of the robot environment. Because the actual program is way too long, only an example to achieve a smile action is demonstrated here:

```
#include"SoftwareSerial.h"

void setup()

{

  Serial.begin(9600);

  Serial1.begin(9600);

}

void loop()

{

  switch(Serial.read())

    {

        case 20:

                {

                  Serial1.print("G21F3\r\n");

                  delay(800);

                }break;

        default:  break;

    }

}
```

In case 20, 20 is the return value for the voice module receiving a smile command transferred to the microcontroller. 21 in the serial output of G21F3 is the action group label of smile and 3 is the number of executions. If you want to achieve more actions, it is only needed to add action group and set the corresponding return value.

6. Conclusions

Through the study of multi-function robots with speech interaction and emotion feedback, combined with the Internet of things technology, not only the basic functions of the case are realized, but also the self-help hotel applications are completed with the aid of the robots. In conjunction with the increasing demand for service-oriented and networking products in the today's society, we believe that this robot will certainly have a broader development prospects in the future.

7. Acknowledgement
The project is supported in part by the Undergraduate Innovative Discipline Project in China (201610359048) and the National Science Foundation of China (51405126).

8. References
[1] Qin D, 2003 Alive robots, *News Lover*, April 30(46)

[2] Liu R, Peng S and Liu G, 2014 Speech recognition embedded in intelligent house-holding system, *Journal of Guangdong University of Technology*, 2, pp. 50-53

[3] Li H, Mao X and Song J. 2016 Intelligent bionic two-way sign language translation system, *Application of Electronic Technology*, 42(7), pp. 83-86

[4] Su B, 2012 Design of speech recognition system based on AVR microcontroller, *Modern Electronic Technique*, 35(11), pp. 136-138

[5] Chen S, Wang Q and Wang T, 2013 A data transmission system based on W5100, *Application of Electronic Technology*, 39(9), pp. 54-60

[6] Ji X, 2012 Design of intelligent light-seeking car based on Arduino development environment, *Modern Electronic Technique*, 35(15), pp. 161-163

[7] Wu Y, Li L, Chen S and Pen W, 2014 Design of photoelectric encoder detector based on Arduino development environment, *Modern Electronic Technique*, 37(2), pp. 124-125

[8] Cui Y, Zhang W and Bai Y, 2013 An intelligent home control system based on Arduino, *Application of Electronic Technology*, 39(9), pp. 54-60

Research on Rigid Body Motion Tracing in Space based on NX MCD

Junjie Wang[1], Chunxiang Dai[1] , Karen Shi[2] and Rongkang Qin[1]

[1] Mechanical Manufacturing and Automation Institute, Shanghai University, No.149 YanChang Road, Shanghai, China
[2] SIEMENS Product Lifecycle Management Software Inc., No.1018 ChangNing Road, Shanghai, China

Abstract. In the use of MCD (Mechatronics Concept Designer) which is a module belong to SIEMENS Ltd industrial design software UG (Unigraphics NX), user can define rigid body and kinematic joint to make objects move according to the existing plan in simulation. At this stage, user may have the desire to see the path of some points in the moving object intuitively. In response to this requirement, this paper will compute the pose through the transformation matrix which can be available from the solver engine, and then fit these sampling points through B-spline curve. Meanwhile, combined with the actual constraints of rigid bodies, the traditional equal interval sampling strategy was optimized. The result shown that this method could satisfy the demand and make up for the deficiency in traditional sampling method. User can still edit and model on this 3D curve. Expected result has been achieved.
Keywords. MCD; simulation; path; sampling strategy.

1. Introduction

In 1970s, the concept of Mechatronics was formally introduced, many design ideas have been applied to this. And Conceptual Design is a process from simple to precise, fuzzy to clear, combining Mechatronics and Conceptual Design can make the project more flexible [1][2]. SIEMENS is involved deeply in the field of Mechatronics Conceptual Design. The research of this paper was based on MCD (Mechatronics Concept Designer) which is a module of SIEMENS industrial design software UG NX. In the design process, consider from completing the task, drive and actuation mechanism is the core of the machine. Therefore, in the simulation of the conceptual model, users are concerned about the path of these two mechanisms [3][4]. This paper will aim at the requirement of tracing the path of the actuator in models, realize it through real time calculation of actuators' position and fitting of these result points. Meanwhile, this paper proposed an optimized sampling method which would first analyze the restrained condition of actuators. Experimental results show that this method can fit well to the path and make up for the deficiency in traditional sampling method.

2. Method

2.1 Transform Matrix

In computer, graphics are processed in the form of numbers [5]. To change the position of a point, a commonly used method is to multiply the coordinate value with a 4×4 transformation matrix. It can make any point move to another position through adjusting the parameters in this matrix. The formula is

IOP Conf. Series: Materials Science and Engineering **320** (2018) 012012 doi:10.1088/1757-899X/320/1/012012

$$(x' \quad y' \quad z' \quad 1) = (x \quad y \quad z \quad 1) \cdot \begin{pmatrix} M_{11} & M_{12} & M_{13} & M_{14} \\ M_{21} & M_{22} & M_{23} & M_{24} \\ M_{31} & M_{32} & M_{33} & M_{34} \\ M_{41} & M_{42} & M_{43} & M_{44} \end{pmatrix} \quad (1)$$

Let $M_{rot} = \begin{pmatrix} M_{11} & M_{12} & M_{13} \\ M_{21} & M_{22} & M_{23} \\ M_{31} & M_{32} & M_{33} \end{pmatrix}$.

The 3×3 matrix M_{rot} controls the basic transformation of the proportion, symmetry and rotation in 3D graphics. The physics engine will convert the position or speed constraint into force and torque, and decompose equivalently into direction of each axis. In this way, the rotation matrix can be obtained in real time calculation. Thus, the matrix can be divided into

$$M_{rot} = M_z(\varphi) M_y(\gamma) M_x(\theta) .$$

$M_a(\delta)$ represents the rotation of δ degrees around the ***a*** axis.

$(M_{41} \quad M_{42} \quad M_{43})$ controls the transformation along the coordinate axis. The solution method is similar to rotation matrix.
However, in MCD kinematics simulation environment [6], it does not involve the transformation of size, perspective, and graphics scale. So in the process of calculation, $(M_{14} \quad M_{24} \quad M_{34})^T$ which controls the transformation of perspective equals to $(0 \quad 0 \quad 0)^T$. And M_{44} which controls the size and scale transformation equals to 1.

2.2 Theoretical Derivation & Formula

The pose change of a rigid body in space can be decomposed into translational motion and rotation. In practical engineering problems, two characteristics of rigid body are often adopted. First, and equal in translation. Second, the displacement, velocity and acceleration of each mass element in a rigid body are the same [7].

In this paper, the translational motion is simplified as the coordinate change of the mass point M of this rigid body. And the rotation is simplified as the change of the vector $\overrightarrow{MA}$ which is the difference between M and selected point A which is belong to this rigid body.
In the implementation, each solution achieved by engine called Step, and each step would generate a corresponding 4×4 matrix. Noted that some elements in the matrix equal to 0 or 1, so it was better to calculate the coordinates through motion synthesis instead of multiplying them directly.
Set coordinate value of the sampling point A as (x y z), and the mass center M of this rigid body as $(m_x \; m_y \; m_z)$, point O is the origin. After the rigid body starts to move, the position will change. Now we will set the mass point M_i as $(m_{xi} \; m_{yi} \; m_{zi})$, and set the point A_i (i means step i) as $(m_{xi} \; m_{yi} \; m_{zi})$. Then the formula for A_i is

$$\overrightarrow{OA_\iota} = \overrightarrow{OM_\iota} + \overrightarrow{M_\iota A_\iota} \quad (2)$$

$$\text{Means } \overrightarrow{OA_\iota} = \overrightarrow{OM_\iota} + \overrightarrow{MA} \cdot \begin{pmatrix} M_{11} & M_{12} & M_{13} \\ M_{21} & M_{22} & M_{23} \\ M_{31} & M_{32} & M_{33} \end{pmatrix}.$$

$$\text{The result is } \begin{pmatrix} x_i \\ y_i \\ z_i \end{pmatrix} = \begin{pmatrix} m_{xi} + M_{11} \cdot (x - m_x) + M_{21} \cdot (y - m_y) + M_{31} \cdot (z - m_z) \\ m_{yi} + M_{12} \cdot (x - m_x) + M_{22} \cdot (y - m_y) + M_{32} \cdot (z - m_z) \\ m_{zi} + M_{13} \cdot (x - m_x) + M_{23} \cdot (y - m_y) + M_{33} \cdot (z - m_z) \end{pmatrix}. \quad (3)$$

In this formula, $\begin{pmatrix} m_{x_i} \\ m_{y_i} \\ m_{z_i} \end{pmatrix} = \begin{pmatrix} m_{x_{i-1}} + M_{41} \\ m_{y_{i-1}} + M_{42} \\ m_{z_{i-1}} + M_{43} \end{pmatrix}$.

2.3 User Interface & Data Structure

User interface is shown in the figure 1, and the parameters are shown in the Table 1.

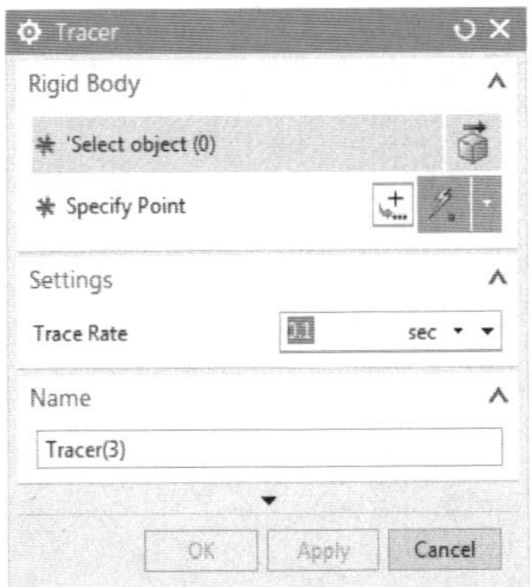

Figure 1. User Interface of Tracer.

Table 1. Parameters in User Interface.

Variable Name	Type	Specification
Select Object	rigid_body	Select a geometry as rigid body.
Specify Point	PNT_3	A point belong to this rigid body.
Tracer Rate	double	Set a rate (sec) of the trace.
Name	string	Set tracer's name.

After sampling an existing rigid body and specifying a point, the tracer will sample this point at intervals. And the trace rate can be also set by user. Then, the result will be calculated by (2).

2.4 Description & Analysis of the Fitting Strategy
A series of coordinate points could be obtained by the above calculation. In order to get the smooth curve, B-spline curve interpolation [8] is used to fit the sampling points [9][10].
Determining a spline curve requires the control point P, then in drawing, the points on the curve will be computed by recursion on the control polygon. For example, a set of points $C = \{c_j : j = 1,2, \ldots, k\}$ is on a B-spline curve, so it should satisfied the definition of B-curve [11][12], which is

$$c_j(u_j) = \sum_{j=0}^{n} N_{j,p}(u_j) P_j, \; j = 1,2,\ldots,k \cdot \tag{4}$$

The formula (4) can be written in matrix form, i.e. $C = NP$. P is a $n \times 3$ matrix with n control points unknown. C is a $k \times 3$ matrix with k 3D points. N is a $k \times n$ coefficient matrix which is the basis functions of the B-spline. If we input a series of points and determine the value of the node vector u, as a result, N can be calculated by de Boor's algorithm [11]. And finally, we can get the matrix P which will include the control points.

The application of B-spline in this paper would meet two special cases. First, the sampling points may lie on the same line in a rigid body motion. One of the properties of B-spline is convex hull property, it makes sure that B-spline should locate in the control polygon surrounding by the nodes. So it can make B-spline degenerate into a line if the control points lie on one line.
The other one is for filtration to multiple-knot. Rigid body may not move in a period of time, if using equal interval sampling, repeat points will be collected at this position. When generating, multiple same points will make the differentiability of the B-spline lower [12], that means will produce "sharp points". The path mentioned in this paper does not need to consider multiple-knot, so it should avoids the occurrence of this situation. Following conditions should be added after each calculation.

```
PNT_3 last_point = samplePointVec.back();  //get the last sample point
if ( last_point.distance (result_point) > 0.01)  //the distance between result point and last point should more than 0.01
    samplePointVec.push_back (result_point);  //insert this point
```

3. Optimization

3.1 Preparation for Optimizing
As the most basic sampling method, Equal Interval Sampling was used in many kinds of literature and also in this paper. Its advantage is that users only need to determine the tracer rate, through condition

$$\text{if}\ (\ (++\,\text{step})\%\ (1000 \times \text{tracer\_rate}) == 0\ )$$

to determine whether matrix calculation should be performed. It has less computation frequency and less memory consumption. However, it has obvious deficiencies in using just like the figure 2 and figure 3. There is little nuance between the two, but the amount of calculation is ten times difference.

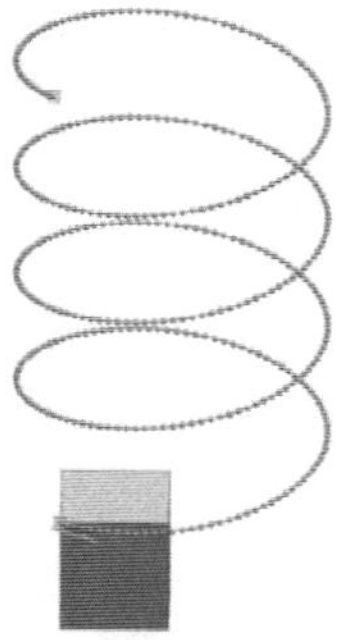

Figure 2. A path line of a block with 0.1ms as trace rate.

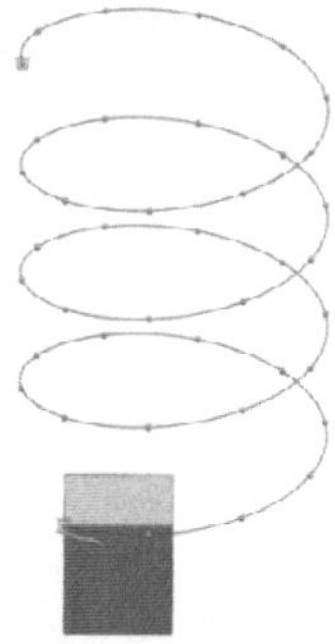

Figure 3. A path line of a block with 1ms as trace rate.

Also in the conceptual design, user can use "Line", "Arc" and "Spline" which are the modelling command of NX to generate curves. Then they can use "Point on Curve Joint" which is the command of MCD to make the rigid body move along this curve. For example, user want to trace point A in the following mechanism shown in figure 4.

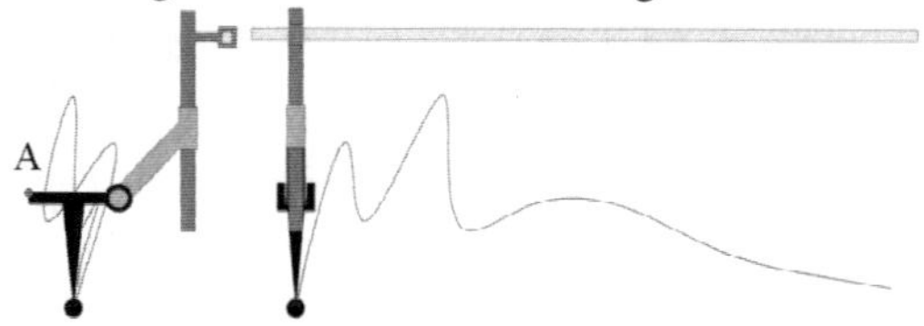

Figure 4. Mechanism for Point on Curve Joint Effects.

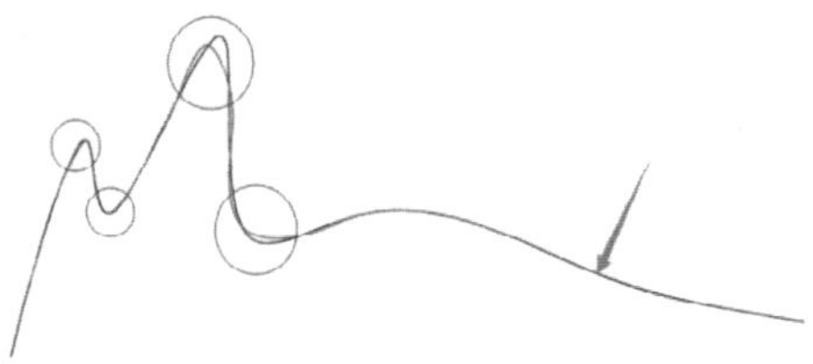

Figure 5. Comparison of Result Curves with Different Sampling Frequencies.
Red: 0.5 & Blue: 0.05

Sampling frequency was set 0.5 and 0.05 respectively, the results have been represented by a red line and a blue line. The figure 5 was the result of comparison.

In contrast, it is obvious that there is a significant fitting distortion in rings. At the arrow point, although great difference in the number of sampling points, the results are similar. In order to save the computational cost and achieve more accurate fitting in tracking, it is necessary to set a higher sampling frequency for the "critical" position of the curve.

Meanwhile in MCD, the joints are referenced on rigid bodies can be analysed. Sliding joint, hinge joint and cylindrical joint will make objects move with linear variation, so Equal Interval Sampling will be a better method for these cases. And when meet "Point on Curve Joint" or "Path Constraint Joint", before sampling, this paper propose that it need to analysis the curve\path to get the critical position. Finally, *Equidistance Sampling Method* can be used for some nonlinear or composite joints situations.

3.2 Analyzing before Sampling

This paper considers that the curvature distribution of a curve can reflect its overall and local characteristics. Therefore, the sampling frequency needs to be related to this curve's curvature distribution. Take figure 6 as an example to illustrate the method. The curve in the graph can be turn into discrete to get points, and the radius of curvature at the ith point can be calculated by the following formula,

$$\rho_i = \frac{|\overrightarrow{d_\iota d_{\iota+1}}||\overrightarrow{d_\iota d_{\iota-1}}||\overrightarrow{d_\iota d_{\iota+1}} - \overrightarrow{d_\iota d_{\iota-1}}|}{2 \times |\overrightarrow{d_\iota d_{\iota+1}} \times \overrightarrow{d_\iota d_{\iota-1}}|}. \tag{5}$$

The value of curvature is small. In order to facilitate calculation and explanation, the results are multiplied by 1000 times, so that it can be drawn in the integer category. The image is shown in the Fig 6,

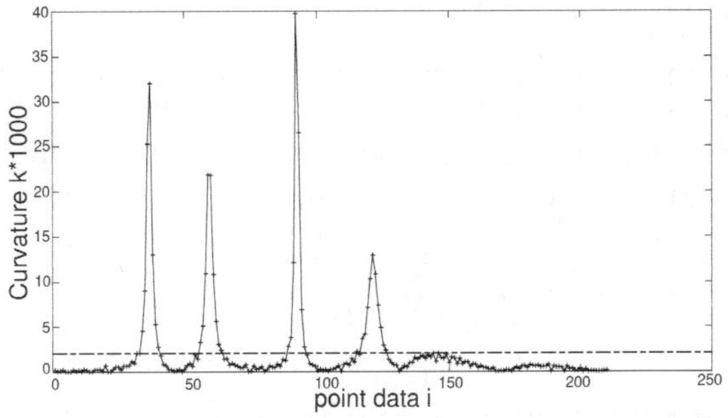

Figure 6. Curvature Distribution.

After the curvature data is obtained, the mean value $\bar{k}$ and overall standard deviations S should be calculated

$$\bar{\mathrm{k}} \approx 2.12, \mathrm{S} = \sqrt{\frac{1}{N}\sum_{i=1}^{N}(Xi - \bar{X})^2} \approx 5.11\,.$$

To determine the critical position, this paper designed a value K′. If it is larger than K′, it represents this section as the key position. In order to maximize the response of key points and also can filter out the others, let $\mathrm{K}' = [\bar{\mathrm{k}}]$, which means integer to $\bar{\mathrm{k}}$.

The curvatures in the following four intervals C1[32, 40], C2[56, 64], C3[89, 95] and C4[115,126] are larger than K'. Their overall standard deviations are S1≈10.300, S2≈7.266, S3≈13.252 and S4≈3.614. And The curvatures in the following five intervals C_5[1, 31], C_6[41, 55], C_7[65, 88], C_8[96, 114] and C_9[126, 210] are smaller than K'. Their overall standard deviations are S_5≈0.294, S_6≈0.573, S_7≈ 0.400, S8≈0.480, S9≈0.559.

Since the overall standard deviation represents the degree of dispersion between data and mean, the greater the degree of discretization, the greater the variation of the curvatures. S1 to S4 are greater than 1, while the remaining S5 to S9 are less than 1, take S as a reference, this paper held that the intervals which curvature value is greater than K′ and S_i is greater than 1 are the critical position. When the rigid body moves to the critical position, the sampling frequency set in this paper needs to refer to the maximum curvature of this interval,

$$R_i = \frac{1}{1000\times(k_{\max})_{ci}} = \frac{(\rho_{\min})_{ci}}{1000},\quad R_{\min} \geq 0.001 \tag{6}$$

After a number of curve tests, the value of S should be less than 10, if greater than, it should not take the above method.

3.3 Equidistance Sampling Method

For the above conditions are not applicable and some complex cases, motion would be more inscrutability. The paper held that the amount of computation needed to be increased to pursue accuracy, so Equidistance Sampling Method performed better.

In the later stage of conceptual design, the motion of many rigid bodies is composed of complicated kinematic joints, such as a point on a connecting rod of a four-bar linkage, or some points in the manipulator's executing mechanism. While in the classical Curve Splitting algorithm [13], the sharp angle of the curve is used as the key frame, because the change between each sharp angle can be expressed linearly, but this algorithm is very heavy in calculation. Another is through the analysis of the rotation matrix.Since the rotation matrix can be decomposed into Euler Angle, through the analysis of angle change can also achieve the purpose. However, to trace the rotation center of a rotating object for example the center of mass of a drill bit. A large amount of calculation but the final result is of no significance.

$$\delta = 1/k \cdot 2^d \tag{7}$$

k represent the degree of accuracy

sampling result will be better. In practice, once this value is exceeded, this point should be record.

4. Result

It's reasonable to use Hausdorff [14] distance evaluation as the error evaluation. The fitting requirement of the data point cloud is that the distance between the fitted curve and the original data point cloud is within a certain tolerance range. The fitting curve of $P_i(i = 1, \dots, n)$ is $C(t)$, projected P_i onto $C(t)$ to get P'_i. The fitting error and average error are

$$e_i = \| P'_i - P_i \| \tag{8}$$

$$e_i = (\sum_{i=1}^{n} e_i^2)^{1/2}/n. \tag{9}$$

In this paper, the average error was used as the basis of judgment for the method which will analyse before sampling.

Take the mechanism in figure 6 as the experimental object of this method, the result is shown by the following figure 7.

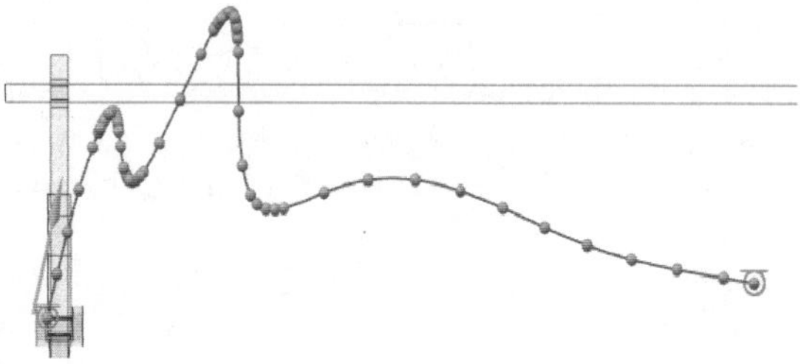

Figure 7. Result of the Method of Analyzing before Sampling.

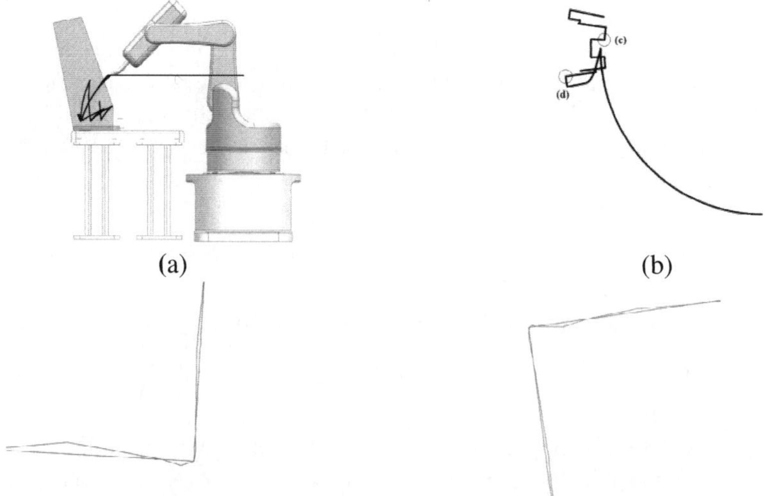

(a) (b)

(c) (d)

Figure 8. Result of the Equidistance Sampling Method.

Compared with equal interval sampling method with 0.03s as the tracking frequency, the optimization method used in this paper can be calculated less than 8400 times in point calculation. The average error between the two paths is 0.05283. When multiple tracers are installed at the same time, more computational costs can be saved.

Then by setting k=1, used *Equidistance Sampling Method* to generate the trajectory of a welding manipulator [15] in figure 10(a) which degree of freedom is 5. And used *Equal Interval Sampling* with the trace rate of 0.05s as contrast.

This paper used red to represent the result of Equal Interval Sampling method, blue to represent Equidistance method. In figure 8 (c) and figure 8 (d), maximum local fitting error of equal sampling method reached 0.164, while equidistance method only 0.009. Also, each cusp in the curve fitted by equidistance sampling method was better than by equal interval.

5. Summary

This paper realized the visual display of rigid body motion trajectory based on Mechatronics Concept Design. And also proposed two sampling optimization methods which are *Analyzing before Sampling Method* and *Equidistance Sampling Method*. The first method will analyse the curvature of constrained curves, determine sampling strategy according to the analysis result. Its advantage is the amount of computation can be reduced substantially with little accuracy loss. However, the curvature is hard to extract. *Equidistance Sampling Method* can give a more accurate result, but the amount of calculation is much larger than the other method. In the future research, how to combine the strategy of sampling and fitting more effectively can be a research direction.

6. References

[1] Yinli Jin, Jianpin Zhang and Kun Ye. 2013 Energy and Energy Conservation. *Parametric Modeling of Large Fan Blades based on NX Secondary Development.* chapter 6, pp. 133-135

[2] Jack Xiong, Chunxiang Dai and Karen Shi. 2016 Metrology & Measurement Technique. *Research and Application of NX mechatronic Concept Design System.* chapter 12, pp. 9-11

[3] A. Roennau, F. Sutter, G. Heppner, J. Oberlaender and R. Dillmann. 2013 16th International Conference on Advanced Robotics. *Evaluation of Physics Engines for Robotic Simulations with a Special Focus on the Dynamics of Walking Robots.* pp. 1-7

[4] Jianshe Guo, Biaoguang Sun and Yunpeng Zhang. 2011 Machinery. *Manual Mechanical Simulation of Punch Machine based on UG.* chapter 21, pp. 479

[5] Di Wu and Wenqian Huang. 2003 Hydrographic Surveying and Charting. *Deduction of Transformation Matrix of Perspective Projection in Graph Transformation.* chapter 23, pp. 18

[6] *PhysX by NVIDIA.* in Ver. 3.1, more details at: http://www.geforce.com/hardware/technology/ physx website 2013.

[7] L. Glondu, M. Marchal, G. 2010 ASME Conference Proceedings. *Dumont, Evaluation of Physical Simulation Libraries for Haptic Rendering of Contacts between Rigid Bodies.*

[8] Rama Krishna N Parasaram and T N Charyulu. 2012 International Journal of Mechanical Engineering and Robotics Research. *Airfoil Profile Design by Reverse Engineering Bezier Curve.* Vol.1, No.3, pp. 410-420

[9] Yoshimoto F, Harada T, Yoshimoto Y. 2003 Computer-Aided Design. *Data fitting with a spline using a real-coded genetic algorithm.* chapter 35(8), pp 751-760.

[10] Kenneth Renny Simba, Naoki Uchiyama, and Shigenori Sano. 2016 Robotics and Computer-Integrated Manufacturing *Realtime smooth trajectory generation for nonholonomic mobile robots using bezier curves.* chapter 41, pp 31–42.

[11] *B-spline Definition & Introduction.* details and manual at: https://en.wikipedia.org/wiki/B-spline, website 2017.

[12] HAN Jiang, JIANG Ben-chi. 2010 Applied Mathematics and Mechanics. *A B-Spline Curve Fitting Algorithm Based on Contour Key Points.* chapter 36, pp. 423-431

[13] Hongmei Zhou, Yanming Wang. 2008 Journal of Xi'an Jiaotong University. *Fitting of non-uniform Rational B-spline Curves based on Minimum Control Points.* chapter 42, pp. 73-77

[14] Park H. 2004 Computer Aided Geometric Design. *An error-bounded approximate method for representing planar curves in B-splines.* chapter 1, pp. 9-12

[15] Harshal B Sonawane and Eknath R Deore. 2014 International Journal of Mechanical Engineering and Robotics Research. *Finite Element Model for the Effect of Heat Input & Speed on Residual Stress during Weldings.* Vol.3, No.3, July 2014

IOP Conf. Series: Materials Science and Engineering **320** (2018) 012013 doi:10.1088/1757-899X/320/1/012013

An Optimization of Manufacturing Systems using a Feedback Control Scheduling Model

John M Ikome[1] and Grace M Kanakana[2]

[1]Department of Industrial Engineering & Operation Management, Vaal University of Technology, South Africa
[2]Department of Industrial Engineering, Tshwane University of Technology, Pretoria, South Africa
Email: Ikome20022000@yahoo.com; kanakanamg@tut.ac.za

Abstract. In complex production system that involves multiple process, unplanned disruption often turn to make the entire production system vulnerable to a number of problems which leads to customer's dissatisfaction. However, this problem has been an ongoing problem that requires a research and methods to streamline the entire process or develop a model that will address it, in contrast to this, we have developed a feedback scheduling model that can minimize some of this problem and after a number of experiment, it shows that some of this problems can be eliminated if the correct remedial actions are implemented on time.

1. Introduction

Now our days, manufacturing companies generate and update production scheduling which is a very critical part in the production process in converting raw materials into useful output. Scheduling allocates the start and finishing time to each particular order of jobs and by also bringing productive operations in the shop floor which actually provide a calendar of step after step for processing a set of jobs using machine, and all other required resources. In the real world situation where we have very complex manufacturing systems, operation managers, line managers and supervisors must not only generate high-quality schedules but also react quickly to unforeseen disturbance that upset the entire production process, in the most economical way that will benefit the company and the customers. This unexpected disturbance are generally difficult to anticipate during the production process, so there is a need to develop a strategic feedback re-scheduling model that can streamline and accommodate this unforeseen disturbance in order to prevent the system to degenerate to a complete short down or defects to products, causing a complete disaster to the economic well-being of the company.

Actually in real practice, production scheduling is a complex system within the manufacturing process that has the flow of information, processes and decision-making that forms part of the manufacturing planning and control system.

In recent years, more attention has been given to the uncertainties while designing schedule models but in the manufacturing shop floor, is practically impossible to know a period or the time when the system can be subjected to a disruption.

1.1. Scheduling

According to Katrina et al, [1] this scheduling problem has commonly be defined in academic literature with emphasizes only on a limited part of the scheduling task. From the aforementioned, following the exact scheduling in a real manufacturing facility as it was designed, is practically impossible due to the difficulties in having an accurate estimation of time estimation for data processing and also the unforeseen disruptions that affects the system.
In justification of the aforementioned, a recent research carried out by Pinedo [2] enumerates a number of difference between theoretical models and real-life production scheduling problems which includes;
The constant arrivals of new jobs into the system,
Processing time not following the statistical distribution,
Complicated machine environment etc.
He remarks that the emphasis of rescheduling problem is not sufficient in the literature of production scheduling models and concluded that when designing a scheduling system, uncertainties need to be taken into account, since as production schedules becomes inaccurate or infeasible as time goes on, requiring a new schedule.
With respect to rescheduling model and approached found in research work, deals with mathematical models in order to optimize the production system, Dutta et al [3], Davenport, [4]. However, these approaches and their results remain well-grounded because the can only be applicable to specific problems and cannot be applicable with complex problems. According to Aytuy and Mchay [5] simulation has been employed in a number of papers in other to validate proposed method of rescheduling and algorithms but this methods also remain grounded because the results needs to be interpreted according to the simulation system being used .
In view of this limitations, our main objective in this paper is to come up with a supply chain re-scheduling model as a feedback control system that can minimize the effects of this disturbances in the real manufacturing environments were the system is complex due to a number of unforeseen situations. In a similar research work, Vieira *et al.* [6] also presented a rescheduling framework that can be used for classifying and describing rescheduling environments, policies, strategies, and methods. According to his findings, production scheduling theory has had limited impact on real life practice because most scheduling results do not consider important characteristics of the environment in which scheduling occurs. From the aforementioned, it is evidence that researchers have not considered the full dynamical aspects of the manufacturing system, including the planning and control systems.
The relationship between the theory and practice of scheduling was also later discussed by McKay and Wiers [7] here, they describe three principles that explain a practical production scheduling system which includes, (generates partial solutions for partial problems, anticipating and reactions for disruptions, and also being very sensitive in adjusting the time in production situation). .
In this paper, we proposed a conceptualized feedback rescheduling model which will be able to control and take care or minimize the effect of these disturbances that makes the production system and activities to become vulnerable.

2. Production Systems

Production/operations management is a process, which combines various activities, systems and resource to transform raw materials into useful good or service. This activities take place in a controlled method and environment. These set of interrelated management activities which involves input require product/service having the requisite quality level, and the set of interrelated management activities, which are involved in product manufacturing called production management, while using the same concept in the service management, is called operation management Vollmann et.al [8].
All these set of activities and operation, need to be done in an orderly manner in order to achieve the key objectives of any manufacturing or service industry which is productivity out-put in order for the company to make money, attract more customs and expand globally.

2.1 Customers service

IOP Conf. Series: Materials Science and Engineering **320** (2018) 012013 doi:10.1088/1757-899X/320/1/012013

The main objective or initial principle of a systems operation is resource utilization which will satisfy the customer's wants and also enable the organization to maximize its out-put. From the aforesaid, it shows that the customer service is a vital and strategic importance of operations management. The organization including all the various operating systems and method has an obligation to provide product that exceeds the customer's specification and satisfaction both in terms of cost, quality and timing. Thus, making sure that the 'right product and process are being done at the right time, sequence and price' can help to gratify the customer's primary objective.
A number of this aspects are described in four functions in Table 1 which includes, the primary sources of customer satisfaction and must actually reflect as the main objective for operations managers during their production process. According to Aytug, H., M. and G .F Verisel [9], An organization will generally aim reliably and consistently to achieve a specific standards which the operation managers and shop-floor teams have put together, but due to the complex nature of manufacturing process and other influential situations, attempting to achieve these standards becomes almost practically impossible.
This further curses a confusion that influences the operations manager's decisions to a different plan of action to aid in achieving the required customer service, but due to unforeseen disturbance in real operation on the shop-flow, obtaining this new objectives or action plans sometimes become practically impossible and the need to be a backup system or methods in place to assist should in case the need arise.

Table 1. Customer's specifications and objectives sections.

The customer service objective	The resource utilizations objective
i.e. toward deliver contracted/acceptable levels of customer	
i.e. to attain adequate levels of resource service (and hence customer satisfaction) through the supply of goods and services that meets the right specification, right cost and the right time.	*i.e to attain adequate levels of resource utilizations (or productivity) e.g.* achieving an agreed levels of utilizations of materials, machines and labour

2.2 *Resource Utilization*
The Resources are very vital part of any organization because products are produced through them in order to satisfy customer. In addition to this, customer service must be provided with the achievement of effective operations through efficient use of resources. Inefficient use of resources or inadequate customer service leads to commercial failure of an operating system. Operations management is concerned essentially with the utilization of resources, *i.e.* obtaining maximum effect from resources or minimizing their loss, underutilization or waste. The extent of the utilization of the resources' potential might be expressed in terms of the proportion of available time used or occupied, space utilization, levels of activity, etc. Each measure indicates the extent to which the potential or capacity of such resources is utilized. This is referred as the objective of resource utilization.

2.3 *Operation managers*
Operation management is concerned with the achievement of both satisfactory customer service and resource utilization. An improvement in one will often give rise to deterioration in the other. Often both cannot be maximized, and hence a satisfactory performance must be achieved on both objectives.

All the activities of operations management must be tackled with these two objectives in mind. This process are illustrated in figure 1. All this activities have to follow each other and should in case anyone is lacking within this scheduled process, then the entire system may become vulnerable to varieties of disturbance which is then transferred to the customer in the forms of: (late delivery, low quality product, and higher cost).

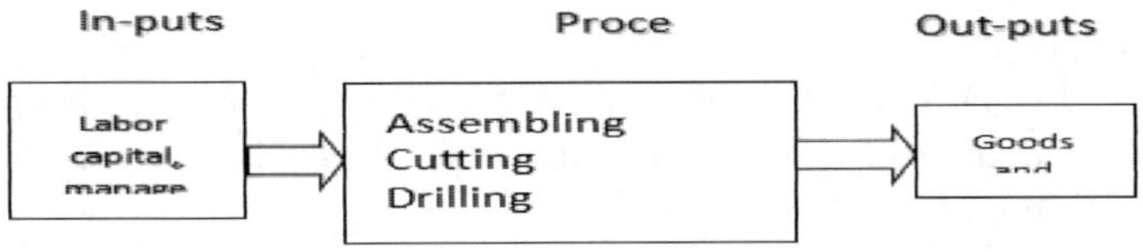

Figure 1. Production system.

According to Ikome et. al. [10], Manufacturing facilities are complex, dynamic and stochastic in nature. Manufacturing systems usually have a wide variety of products and processes and this makes manufacturing managers to count or relay purely on the production schedules because it has a better coordination to increase productivity and minimize operational costs. It can identify resource conflicts, control the release of jobs to the shop, and ensure that required raw materials are ordered in time. In addition, it can also determine whether delivery promises can be met using a number of methods e.g processing methods.

3. Methodology

According to manufacturing and engineering principles, it is evidence that the production process is supposed to flow in an orderly sequence as shown in figure 1, because there are a number of activities and systems that need to perform a number of functions before the optimal goal could be achieved.

In this work, we have developed a feedback rescheduling model that is shown in figure2, which tries to minimize the effect of disruption. A case study company is used to test the model, by combining it with a number of equations to see the various relations. Due to confidentiality, the name of the company will not be mentioned.

Scheduling is the allocation of starts and finish time to each particular order. Therefore productivity improvement in a shop floor can be achieved through a proper scheduling which providing a calendar for processing a set of jobs. We are using a single machine to test the scheduling model, which consists of *n* jobs with the same single operation on each of the jobs, while the flow shop-scheduling consists of *n* jobs with *m* operations on each of the jobs. In this test, all the jobs will have the same process sequences. The job shop scheduling contains *n* jobs with *m* operations on each of the jobs; but, in this case, the process sequences of the jobs will be different from each other.

3.1 Model Concept using a single machine

The basic single machine scheduling problem is characterized by the following conditions:

A number or set of independent jobs are available for a single-operation with a processing time of zero.

Set-up time of each job is independent of its position in jobs sequence. So, the set-up time of each job can be included in its processing time and the Job descriptors are known in advance.

One machine is continuously available and is never kept idle when there is a disruption within the production line or when work is waiting.

Each job is processed till its completion without break.

Under these conditions, the correspondence of one-to-one between a sequence of the *n* jobs and a variation of the job indices 1, 2, ... *n is very visible and straight forward.* The total number of sequences in the basic single machine is *n!* which is the number of different variation of *n* elements.

3.2 Characteristics of identifying problems during scheduling

The following five data are necessary in order to describe or determine any problem during the scheduling process of the model;

Processing time (*tj*): It is the time required to process job *j*. The processing time, *tj* does include both actual processing time and set-up time.

Ready time (*rj*): It is the time at which job *j* is available for processing. The ready time of the job is the difference between the arrival time of that job and the time at which that job is taken for processing. In this model, we condition I, *rj* = 0 for all jobs.

Due date (*dj*): It is the time at which the job *j* is to be completed.

Completion time (*Cj*): It is the time at which the job *j* is completed in a sequence. Performance measures for evaluating this scheduling model is a function of job completion time, and includes, (Flow time, Lateness, Tardiness, etc).

Flow time (*Fj*): It is the amount of time job *j* spends in the system, and it indicates the waiting time of jobs in the system, this will give us some idea about the time it take to process the jobs on the system or work-in-progress (WIP).

Lateness (*Lj*): It is the amount of time by which the completion time of job *j* differs from the due date (*Lj* = *Cj* – *d*). *This will be use as a* measure or idea about conformity of the jobs in the schedule model, to a given period of time or set of due dates of the jobs. This will either be positive lateness or negative lateness.

Meaning the job is completed after its due date or before its due date. The positive lateness means poor service while negative lateness means better service. Note that, in many situations, distinct penalties and other costs are associated with positive lateness. So we are trying to minimize this using our model in order to prevent the company from facing this consequences.

The following equation had been derivate and the different measures of performance which are used in the scheduling model are listed below:

$$Tj = \max\{O, Cj - dj\} = \text{Max}\{O, Lj\}. \quad (1)$$

Mean flow time: F = 1/n

Mean flow time: $$\bar{\mathrm{F}} = 1/n \sum_{j=1}^{n} (F_j) \quad (2)$$

Mean tardiness: $$\bar{\mathrm{T}} = 1/n \sum_{j=1}^{n} (T_j) \quad (3)$$

Maximum flow time: $$F_{max} = Max_{1 \le j \le n}^{Fj} \quad (4)$$

Maximum tardiness: $$T_{max} = Max_{1 \le j \le n}^{Tj} \quad (5)$$

Number of tardy jobs: $$N_T = \sum_{j=1}^{n} f\binom{n}{k} \quad (6)$$

Where: $f(T_j) = I$, if $T_j > 0$, and $f(T_j)=0$

According to Ikome et. al. [11] , in the manufacturing facilities, *production scheduling are very vital part of the organization as it* is a dynamic network or system that share information about the manufacturing process in order to make decisions about which jobs should be done first. These information includes, (job status, work orders, manufacturing resources, raw materials, work-in-process, people and production lines etc.

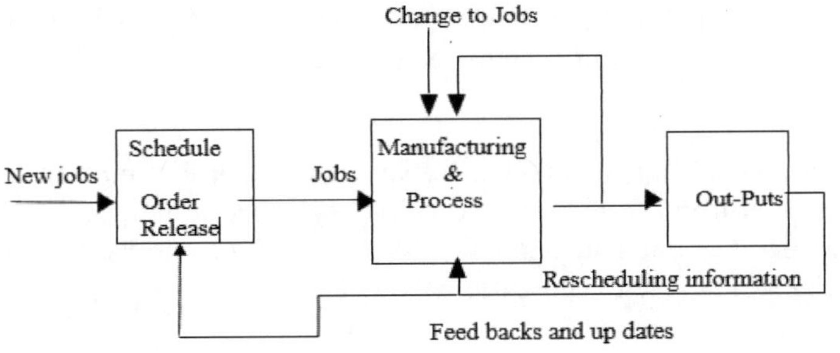

Figure 2. Production Scheduling Model as a feedback Control System.

4. Case Study and Testing Model

In this model, there are a number of jobs that have to be process. During the production process, there is a continuous update that informs the system about the Schedule status and the total number of jobs on the system. By so doing, it prevent raw material from flowing to the downstream should in case there is a disruption with the work-in-progress. This method minimizes defect and waste by not allowing more material to flow to the shop floor until the disruption is addressed. In an attempt to further strengthen this model, a backup generator is included on the system to eliminate the possibility of disruption due to power failure. In addition, there is also an emphasis for the employees to be cross train, this will help prevent the issue of zero productivity due to employee's absenteeism.

The following are among the key decision areas in the feedback scheduling and control model that are taken into consideration:

- Releasing jobs for production,
- Power failure
- Determining when jobs should be started, and
- Assigning resources (people, equipment, or production lines) to jobs,
- Disrupted jobs that should be halted.
- Reassigning resources from one job to another,
- Employee absenteeism
- Unskilled employees.
- Prioritizing jobs that require the same resources,

Note that, after the schedule is generated, manufacturing operations begin and our objective is to see the shop floor following the schedule and also reacting during the unforeseen disturbance. In real practice the operators in the shop floor may deviate partially from the schedule. Preferably, the schedule should be followed as closely as possible. A number of Minor deviations which includes, (Minimum start and end times) are inevitable and are ignored.

According to Vieira *et al.* [12], Rescheduling is a key principle for understanding production scheduling systems. *Rescheduling* is the process of updating an existing production schedule in response to disruptions or other changes which includes different types of disturbances that can upset a production schedule, namely machine failures, processing time delays, rush orders, quality problems, and unavailable material. In our experiment, the rescheduling is done periodically in order to plan

IOP Conf. Series: Materials Science and Engineering **320** (2018) 012013 doi:10.1088/1757-899X/320/1/012013

activities for the next time period based on the state of the system and also occasionally in response to the magnitude of the disruptions.
However, although this model looks simple, note that production scheduling systems are complex in nature because the mechanisms for sensing the state of the manufacturing system and generating updated production schedules cannot be expressed in detailed (except in very special cases) as mathematical functions. In addition, the randomness of disruptions and other uncertainties makes it difficult. The most realistic representation will be to view the systems as a decision-makers.

4.1 Model evaluation
In order to evaluate the model, an automotive gasket manufacturing company was taken as an applicable example for experiment. In the production line, considering it series set-up, there are five different work stations that perform a number of activities, each of this work station comprises of several machines, which execute functionally different process.

5. Results
We considered a single machine production line that runs in series to do our experiment, sequencing the jobs in increasing order of processing time which is known as the shortest processing time (SPT) sequencing. We were mostly interested in minimizing the time spent by jobs in the system in order to minimize the work-in-process inventory. We also, had to pay some interested in the rapid turnaround/throughput times of the jobs.
One of our main objective was to find the optimal sequence, which will minimize the mean flow time and also obtain the minimum mean flow time after the system is disrupted due to unforeseen circumstances. The total number of jobs under this situation is five. We had to arrange the jobs in an orderly manner of Shortest Processing Time (SPT) please see table1, and fig 3. for results

Table 2. No of Jobs to be process after Disruption.

Job (j)	Processing time/Out-put (t) (hrs)
1	15
2	4
3	5
4	14
5	8

Figure 3. Disrupted flow.

IOP Conf. Series: Materials Science and Engineering **320** (2018) 012013 doi:10.1088/1757-899X/320/1/012013

In order to evaluate the model, a first test was done and the results are presented in table 2 and fig 4. From the results, it show the period when the was a disruption and there is a clear indication that the processing time does increase due to the disruption, requiring the workers to work overtime in order to meet up with the initial out-put set by management, and this is a situation that NO company want to find itself in. Please see the table 3 and the figure 4 that follows for further result after testing the model.

Table 3. Job arrangement according to processing time (SPT)

Job (j)	Processing time/Out-put (t) (hrs)
1	4
2	5
3	8
4	14
5	15

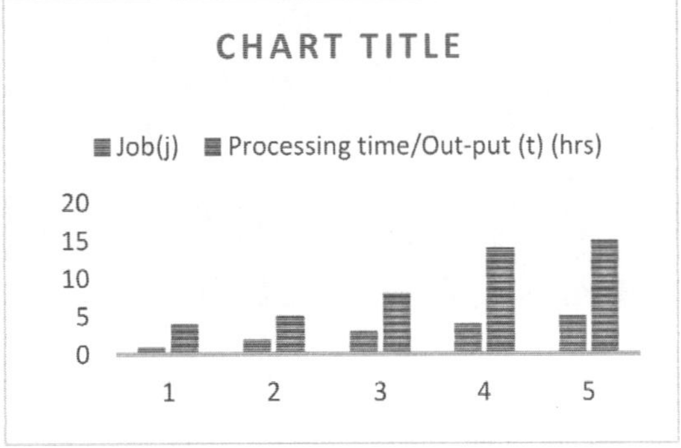

Figure 4. Job sequence and expected out-puts.

By applying equation (2) to minimize the flow time after disruption of the jobs on the system, it show that the jobs sequence, which will minimize the mean flow time, is 2-3-5-4-1 and by computation of Fmin. Figure 3 further show the various processing time and expected out-put. Please see table 3 for results.

In order to further evaluate the feedback scheduling model, the approach of using a dual production line was taken in to consideration. In this duel production line, in total there are five different work piece variants from the raw material to the final product. Table 4, shows how this materials run through multi-level process chain which includes; Power press, bench grinding, and fly press, drilling and dies fixtures.

Table 4. Completion time and Total out-put percentage of jobs

Job(j)	Power Press	Bench Grinder	Fly Press	Drilling	*Dies &Fixtures*
Processing time (t) (hr)	4	5	8	14	*15*
Completion time (Cj) (Fj)	4	9	17	31	*46*
Percentage	45%	53%	25%	68%	18%

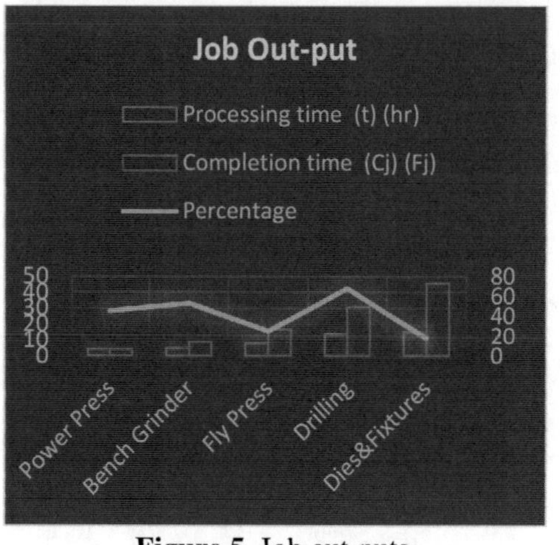

Figure 5. Job out-puts.

Fig.5 shows the final validation of the model, which indicated a high planning quality due to the minimization of disruption effect during the experiment. The real out-put of the production lines shows that after disruption, compared to previous scenarios, it improves the company's average processing time by 19 percent.

6. Conclusion

In this work, a novel approach is presented, which show a number of gap between theoretical and real-life situation in manufacturing industries when disruption emerges, and also how manufacturing systems are very complex in nature. It further compares a number of methods that have been used in order to solve this problem and was unsuccessful because of the complex nature of the system, some were solved and others left out. Our model was developed using some of these unsolved problems and was tested in an automotive gasket manufacturing company, and the result shows that disruption can be minimize if the right steps are taken on time.

The depicted scenario focuses on the scheduling of single and dual production lines in a complex production system with an unlimited number of production orders and machines. In addition, we also anticipated that this research work will help Manufacturing managers and engineers, to improve their production scheduling system.

Future research will focus on a by-pass scheduling system that can automatically rectify any disruption.

7. Acknowledgments

The author will like appreciate the SA Automotive gasket manufacturing industry for granting us permission to their facility in order for the experiment to be conducted and also special thanks to Prof

Maurice Ndege of Vaal University of Technology for encouraging us to do research in the faculty of engineering.

8. Reference

[1]. Katragjini, Ketrina, Eva Vallada, and Rubén Ruiz. 2013. "Flow shop rescheduling under different types of disruption." International Journal of Production Research 51 (3): 780–97. doi:10.1080/00207543.2012.666856.

[2]. Pinedo, M., J. C. Bean, J. R. Birge, J. Mittenthal, and C. E. Noon. 2006. "Matchup scheduling with multiple resources, release dates and disruptions." *Operations Research* 39 (3): 470–483.

[3]. Dutta, A., P. Leitao, and F. Restivo. 2007. "A holonic approach to dynamic manufacturing scheduling." *Robotic and Computer Manufacturing* 24: 625–634.

[4]. Davenport, A. 2000. "Reacting to scheduling exceptions in FMS environments." Transactions 22 (4): 300–14.

[5]. Aytug, H., M. and K. McKay. 2005. "Executing production schedules in the face of uncertainties: A review and some future directions." *European Journal of Operational Research* 161 (1): 86–110. http://dx.doi.org/10.1016/j.ejor.2003.08.027.

[6]. Vieira, G. E., A. K. Jain, and H. A. Elmaraghy, 2007. Production scheduling/rescheduling in flexible manufacturing. *International Journal of Production Research* 35: 281–309.

[7]. McKay, Kenneth N., and Vincent C.S. Wiers, "Unifying the theory and practice of production scheduling,"*Journal of Manufacturing Systems*, Volume 18, Number 4, pages 241-255, 1999.

[8]. Vollmann, Thomas E., William L. Berry, and D. Clay Whybark, *Manufacturing Planning and Control Systems*, fourth edition, Irwin/McGraw-Hill, New York, 1997.

[9]. Aytug, H. M. and G .F Verisel. 2008. "Diruption in production schedules in the face of uncertainties: Case study food manufacturing." *European Journal of Operational Research* 154-632

[10]. John M. Ikome, Sesan P. Ayodeji & Grace M. Kanakana (2016) Minimising instability on manufacturing systems after random disruption, African Journal of Science, Technology, Innovation and Development, 8:2, 142-145, DOI: 10.1080/20421338.2016.1147198.

[11]. J.M. Ikome, M.G. Kanakana & S.P. Ayodeji (2015) The deterioration of production schedules during unforeseen disruptions, African Journal of Science, Technology, Innovation and Development, 7:6, 480-484, DOI: 10.1080/20421338.2015.1097232

[12]. Vieira GV, Jeffrey W H, Eward L. Rescheduling Manufacturing System: A Framework of Strategies, Policies and Methods. Journal of Scheduling 6: 39-62, 2003.

IOP Conf. Series: Materials Science and Engineering **320** (2018) 012014 doi:10.1088/1757-899X/320/1/012014

Adaptive Sampling based 3D Profile Measuring Method for Free-Form Surface

Xianyin Duan[1,2], Yu Zou[1], Qiang Gao[1], Fangyu Peng[2], Min Zhou[3], Guozhang Jiang[3]
[1] Key Laboratory of Metallurgical Equipment and Control Technology,
Wuhan University of Science and Technology, Wuhan, China;
[2] State Key Laboratory of Digital Manufacturing Equipment and Technology,
Hua zhong University of Science and Technology, Wuhan, China;
[3] Hubei Key Laboratory of Mechanical Transmission and Manufacturing Engineering,
Wuhan University of Science and Technology, Wuhan, China.

Abstract. In order to solve the problem of adaptability and scanning efficiency of the current surface profile detection device, a high precision and high efficiency detection approach is proposed for surface contour of free-form surface parts based on self- adaptability. The contact mechanical probe and the non-contact laser probe are synthetically integrated according to the sampling approach of adaptive front-end path detection. First, the front-end path is measured by the non-contact laser probe, and the detection path is planned by the internal algorithm of the measuring instrument. Then a reasonable measurement sampling is completed according to the planned path by the contact mechanical probe. The detection approach can effectively improve the measurement efficiency of the free-form surface contours and can simultaneously detect the surface contours of unknown free-form surfaces with different curvatures and even different rate of curvature. The detection approach proposed in this paper also has important reference value for free-form surface contour detection.

1. Introduction

The surface profile detection device can be used to detect the surface profile of a manufactured mechanical part. As an indispensable instrument in reverse engineering, profile inspection device plays an important role in the mechanical field [1-3]. The current surface profile measurement approach is mainly contact and non-contact two categories [4, 5].

Wang C L et al. designed a measuring system of the cone pipe thread profile based on the principle of laser triangulation. The automatic measurement of thread profile was realized by non-contact scanning approach [6]. Sun Y L et al. analysed the causes of the non-linear error in the process of large-scale contact profile measurement, improved the structure of the measuring bar, and established the nonlinear error model [7]. Zhu D H et al. used the technical reverse engineering to fuse contact measurement technology and non-contact measurement technology [8]. Therefore, the combination of

IOP Conf. Series: Materials Science and Engineering **320** (2018) 012014 doi:10.1088/1757-899X/320/1/012014

contact measurement and non-contact measurement technology is an important approach to realize efficient and high-precision detection of self-adaptive profile and free-form surface parts [9-11].
In this paper, we propose a high efficiency and high precision detection approach for surface profile of free-form surface parts, which based on self-adaptability. Aiming at the issues of self-adaptability, precision, and_efficiency on the free-form surface profile detection, we design a surface profile measurement technique based on contact measurement and non-contact measurement. Through the planning of the scanning path, the detection device can adapt to detect the free-form curved surface, and improve the efficiency and precision.

2. Design of self - adaptive detection device for surface profile of free-form surface parts

The mechanical structure of the surface profile detection device for the free-form surface part designed in this paper can be divided into three parts: supporting parts, driving parts and measuring parts. The support part comprises a base, a workbench and a bridge bracket. The driving part comprises a guide rail, a connecting rod which is CMM (Coordinate Measuring Machine) main shaft, and a rotating table. And the measuring part is composed of a touch sensor and a non-touch sensor.
The detecting device can accomplish the motion of multiple free-form degrees and perform the profile detection for the components of the free-form curved surface. The worktable is equipped with a sliding guide rail, and the whole bridge bracket can be moved along the guide rail. The guide rail is also mounted on the beam, and the CMM main shaft can be moved along the rail. The measuring part is mainly composed of touch-trigger probe and laser probe. The structure of the measuring part and the rotating table are shown in figure 1.

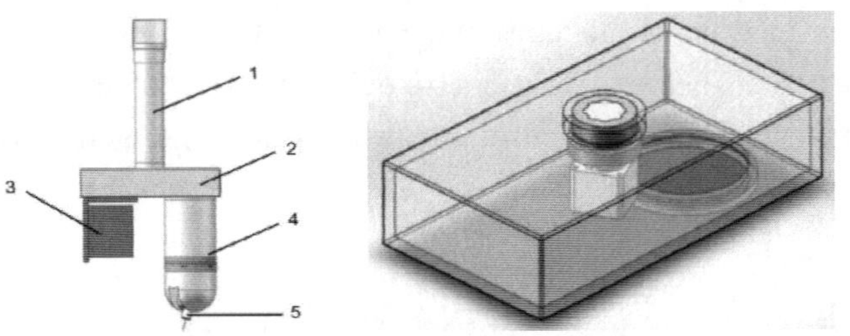

(a) Measuring part (b) Rotating table
1-Telescopic rod, 2-Rotating table, 3-Laser probe,
4-Touch-trigger probe head, 5-Touch-trigger probe
Figure 1. The structure of the measuring part.

3. Scanning process and principle of adaptive profile scanning

As shown in figure 2, the sequential scanning process of the free-form profile surface of the entire part is designed. In the scanning process, the laser sensor (LP) which is in front of the touch-triggered sensor (TP) collects and sorts the data, and pre-judges the sampling frequency of TP. If we only scan the path L_1, we can get the outline of L_1. The concrete steps are as follows:
Step1 After manually determining the initial scan point and the positive direction of the X and Y, the starting point of the probe path L_1 is determined by LP. When LP captures the mutated signal, the coordinate value is recorded, i.e. LP enters the boundary point of the surface. At this point, the servo motor which controls the movement of Y axis is locked. At the same time, the servo motor of the rotary table is locked, and the sensor cannot rotate around the Z axis. Then, the measuring system moves in the

positive direction of the X axis. When TP moves to the boundary point, then re-measurement starts and the result is integrated into the system and recorded as the starting point. If the error between measured value and the coordinate value recorded by LP is greater, the machine alarms, all movement stops.

Step 2 In the process of scanning in the X positive direction, LP has been in the front-end detection, and L_1 is measured at a pre-set unit length ΔX to obtain the coordinate information of the surface profile **point**. When the curvature value $\mathbf{K_i}$ of TP scan section is greater than the pre-set curvature value $\mathbf{K_0}$, the scanning frequency increases. When the curvature value $\mathbf{K_i}$ of TP scan section is smaller than the pre-set curvature value $\mathbf{K_0}$, the scanning frequency reduces. Due to the higher precision of TP and combined with the measurement results of TP, the key points of the rough curve are modified to synthesize the final curve.

Step 3 When LP captures the mutated signal, the coordinate value is recorded. Then, the measuring device continues to move distance d+$\Delta\mathbf{d}$ in the positive direction of X (d is the horizontal distance between the centre axis of the TP rod and the laser beam emitted by LP, and $\Delta\mathbf{d}$ is the pre-set value of the measuring instrument). When TP moved to the boundary point, re-measurement is performed and the results are integrated into the system to synthesize the curve. If the error between measured value and the coordinate value recorded by LP is greater, the machine alarms, all movement stops.

Step 4 The turntable servo motor unlocks and rotated 90° around the Z axis, so that LP is on the left, TP is on the right. Then, the turntable servo motor is locked.

Step 5 The servo motor which controls the movement of the Y axis unlocks and the measuring device move ΔY in the positive direction of the Y axis. Then, the servo motor is locked. At this point, according to whether the focus is on the measuring table, LP judges whether the surface has been entered or not. If so, LP continues to move in the positive direction of the X until the mutation point of the signal is encountered. Then step (6) begins to be executed. If not, the step (6) is executed directly.

Step 6 Repeating steps (1) to (5) to scan the path L_2 until the entire free-form curved surface is scanned.

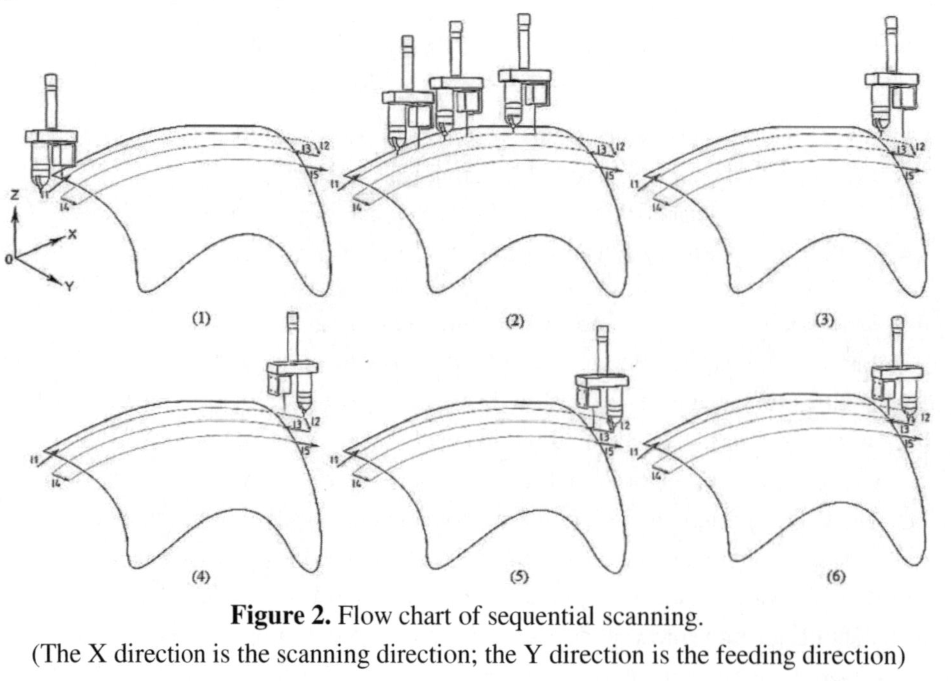

Figure 2. Flow chart of sequential scanning.
(The X direction is the scanning direction; the Y direction is the feeding direction)

4. Information flow and signal processing during scanning

In the adaptive sampling process, the measurement system is responsible for collecting the relevant point information and control system to determine the validity of the point. And it plans the adaptive sampling path. Then the path information is uploaded to the computer to achieve man-machine interaction and control the execution path of the drive system. The protection system is responsible for real-time monitoring, the system will alarm if illegal operation or to reach the movement limit. The flow of information during the scanning is shown in figure 3.

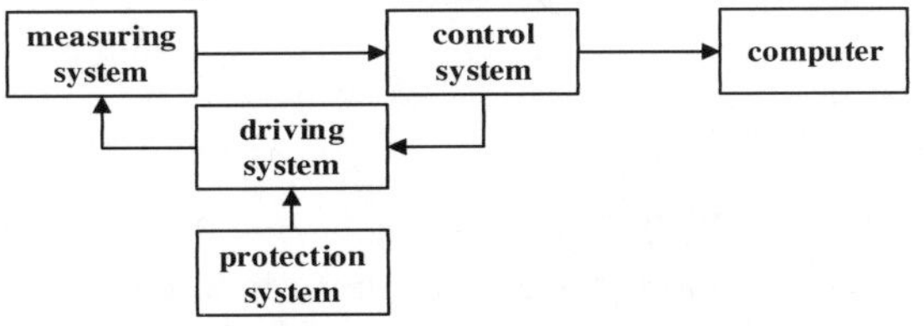

Figure 3. Information flow in scanning process.

The two measuring heads of TP and LP work separately and sample individually at a certain time interval. The signal processing includes the elimination of noise signals, data smoothing, and data compression. TP sampling point instead of LP sampling points and other procedures. The data processing flow of LP is shown in figure 4.

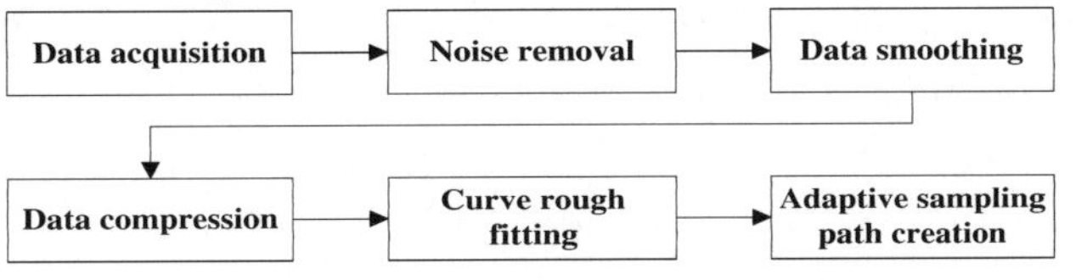

Figure 4. Data processing flow of LP.

Step 1 Data acquisition. In the measurement system, LP collects data and uploads the data to the control system for processing. The sampling interval of data acquisition is $\mathbf{\Delta x}$, which is the setting value by users. The smaller the ΔX is, the higher the sampling precision of LP is.

Step 2 Noise removals. Noise removal is the pre-treatments of control system for the data. The coordinate abnormal points will appear during the scanning process. And in order to improve the final fitting of the rough curve to achieve a certain quality requirements, we need to remove these abnormal points in advance to improve the signal to noise ratio.

Step 3 The traditional signal noise processing is the spectrum analysis. But the data collected by LP is a two-dimensional discrete point, with a certain degree of particularity. We can take a more simple approach of data pre-processing.

Step 4 We use the two-dimensional discrete lattice captured by LP as a separate aggregation. In order to reduce the amount of data processing, only X, Z coordinates points are extracted, that is the aggregation of $\mathbf{G_i(X_i,\ Z_i)}$. When the angle between the line $\mathbf{G_iG_{i-1}}$ and the line $\mathbf{G_iG_{i+1}}$ is smaller than the pre-set angle, $\mathbf{G_i}$ are judged as unusable points, and the points are deleted immediately.

Step 5 Data smoothing processing. The linear smoothing algorithm is used to smooth the available points after noise removal, which is convenient for the final curve creation. The smoothing algorithms used in this paper are as follows:

$$\begin{cases} G_i = (G_{i\ 1} + G_i + G_{i+1})/3 \\ G_0 = (5G_0 + 2G_1 - G_2)/6 \qquad (i = 1, 2, \mathrm{L}\ , m-1) \\ G_m = (-G_{m-2} + 2G_{m-1} + 5G_m)/6 \end{cases}$$

Step 6 Data compression. When the amount of scanning is very large, the computer ability of processing the data will slow down, and we need to compress the digitized data after removing the unusable points.

Step 7 Curve rough fitting. The curve fitting approach is based on the spline curve fitting approach.

Step 8 Creating adaptive sampling path. When the computer obtains the data captured by LP to complete the rough curve fitting, the results will be uploaded to the control system. According to the specific criterion approach, the control system generates decentralized adaptive points, which are the sampling points of TP.

5. Conclusions

In this paper, the adaptive sampling approach based on front-end prediction is proposed to adjust the sampling frequency for the parts system with different surface. Moreover, we propose a two scanning process using the section subdivision method in the feeding direction of the scanning. Furthermore, we initiatively put forward the design of the rotary table in the international, which effectively integrates the laser probe and touch-trigger probe, and add the C axis degrees of free-form.

The results from the design of the adaptive sampling approach based on front-end prediction indicate the following: (1) compared with the current traditional uniform point cloud scanning, the speed of scanning and the precision of surface fitting are greatly improved; (2) the scanning precision is improved in the Y axis direction; (3) the operation flexibility of the new measuring instrument greatly increase and the real value of the measured data is obtained by coordinate compensation mechanism; (4) In the measurement process, the probes can be replaced directly to fit the profile detection of different surfaces.

6. Acknowledgements

This work was supported by National Natural Science Foundation of China under Grant No. 51605346, China Postdoctoral Science Foundation under Grant No. 2016M602374, and the open fund of the key laboratory for metallurgical equipment and control of ministry of education in Wuhan university of science and technology under Grant No. 2015B15.

7. References

[1] Zhao P, Li Y K, and Bai X W 2009 Digital measurement method and its application in reverse engineering [J]. Mechanical Design and Manufacture, (2): 98-100

[2] Yoon, D, Shen, J, Mohanty, P and Kuntz, J 2015 Application of sweeping techniques to reverse engineering. International Journal of Modeling and Simulation, 25(4), 278-284

[3] Hui G, Ning J, and Rui U 2015 Reconstruction of aircraft surface based on reverse engineering. Geomatics and Spatial Information Technology

[4] Chen M and Wei W U 2012 Research on Uncertainty Evaluation of Free-form Surface Profile [J]. Machine Tool & Hydraulics

[5] Ye, S. W., Yang, P., and Peng, Y. F. 2016 A profile measurement method of large aspheric optical surface based on optimal stitching planning. Precision Engineering, 45, 90-97.

[6] Wang C L, Yu D H and Han L G 2016 Application of laser triangular method in the measurement of thread profile of cone tube [J]. Tool Technology, 50 (9): 101-104

[7] Sun Y L, Liang Y H and Chang S P 2015 Nonlinear error analysis and compensation of contact surface profilometer [J]. Measurement Technology, (5): 10-13

[8] Zhu D H 2012 Contact measurement technology and non-contact measurement technology integration [J]. Technology and Market, 19 (5): 171
[9] Yu M, Zhang Y and Li Y 2013 Adaptive sampling method for inspection planning on CMM for free-form surfaces[J]. International Journal of Advanced Manufacturing Technology, 67(9-12)
[10] Lu K, Wang W and Wu Y 2013 An Adaptive Sampling Approach for Digitizing Unknown Free-form Surfaces based on Advanced Path Detecting [J]. Procedia Cirp, 10:216-223.
[11] Morita, S. Y., Yamagata, Y. and Guo J 2014 High speed non-contact on-machine profile measurement for ultrahigh precision machining of large free-form optical surface. Jsat, 58

Improved Extreme Learning Machine based on the Sensitivity Analysis

Licheng Cui[1,2,3], Huawei Zhai[3], Benchao Wang[1] and Zengtang Qu[1]

[1] Liaoning Police College, Dalian, 116036, China
[2] Faculty of electronic information and electrical engineering, Dalian University of Technology, Dalian, 116024, China
[3] Information Science and Technology, Dalian Maritime University, Dalian, 116026, China

Abstract. Extreme learning machine and its improved ones is weak in some points, such as computing complex, learning error and so on. After deeply analyzing, referencing the importance of hidden nodes in SVM, an novel analyzing method of the sensitivity is proposed which meets people's cognitive habits. Based on these, an improved ELM is proposed, it could remove hidden nodes before meeting the learning error, and it can efficiently manage the number of hidden nodes, so as to improve the its performance. After comparing tests, it is better in learning time, accuracy and so on.
Keywords: Sensitivity analysis; SVM; Extreme learning machine; Importance

1. Introduction

Extreme learning machine [1](ELM) is a special algorithm of single-hidden layer feed-forward neural networks (SLFNs), all parameters can be selected randomly except the two, the number of hidden nodes and output weight.

Compared with other algorithms, ELM is simple, training time is shorter, but it is also weak in computational complexity and generation ability. So, it needs to be improved. Literature[2] introduced the wavelet transform and proposed a combination predicting method, it uses the wavelet to decompose the data and then predict, this method does not modified the ELM itself, the performances of ELM is not improved remarkably. Literatures[3, 4] introduced the concept of sensitivity of hidden nodes, sorted each hidden node based on the sensitivity, and finally removed several hidden nodes whose sensitivity are least, these resolved design problem of the structure of SLFNs. Literatures[5-8] referred from the idea of sequential algorithms, proposed variable-structure extreme learning machine, which could be adjust the number of hidden nodes with the new inputting samples.

Referencing the idea of the importance in pruning algorithms, this paper modifies the computing method of the sensitivity, and proposed an improved ELM (IPS-ELM for short), which could remove hidden nodes based on their sensitivity.

2. ELM

Based on the ELM, the output of SLFNs is

$$f_M(x) = \sum_{i=1}^{M} \beta_i G(a_i, b_i, x) \tag{1}$$

where, M is the number of hidden nodes, $\beta_i \in R^m$ is the output weight of the ith hidden node, $G(a_i, b_i, x)$ is the activation function of the ith hidden node, $a_i \in R^n$ is the input weight and $b_i \in R$ is the threshold of the ith hidden node. $x \in R^n$ is the input vector.

For given training samples (x_i, y_i), $i = 1,...,N$, x_i is n×1 input vector, y_i is m×1 expected output vector. So, the real output y_i' is

$$y_i' = \sum_{i=1}^{M} \beta_i G(a_i, b_i, x_i),\ i = 1,2,\cdots N \tag{2}$$

After being transformed, the equation(2) can be deployed as $H\beta = Y'$, β is the output weight, $\beta = [\beta_1 \cdots \beta_M]^T$, Y' is the real output, $Y' = [y_1' \cdots y_N']^T$, H is the output matrix of hidden nodes,

$$H = \begin{bmatrix} G(a_1, b_1, x_1) & \cdots & G(a_M, b_M, x_1) \\ \vdots & \ddots & \vdots \\ G(a_1, b_1, x_N) & \vdots & G(a_M, b_M, x_N) \end{bmatrix} \tag{3}$$

In ELM, the parameters a_i and b_i are generated randomly, and the output weight β needs to be regulated as the new sample inputs, its optimal solution is

$$\beta = H^{\dagger} Y \tag{4}$$

where, Y is the real output, $H^{\dagger}$ is Moore-Penrose generalized inverse, if $rank(H) = M$, the equation(4) could be transformed to

$$\beta = (H^T H)^{-1} H^T Y \tag{5}$$

3. Improve ELM

3.1 The sensitivity of hidden nodes

Literatures[3, 4] gives the concept of the sensitivity, according to the function space, the sensitivity of hidden nodes is the space distance between the output results before and after removing hidden nodes.

$$E(l,i) = \left\| f_{bef} - f_{aft} \right\| = \left| h_{li} \right| \left\| \beta_l \right\| \tag{6}$$

$E(l,i) \in (0,+\infty)$, the larger $E(l,i)$ is, the greater the sensitivity is, otherwise, the less it is. $E(l,i)$ is only based on one sample, and not enough to measure the real state, so it needs large amount of samples, so as to they can fully show their change law and trends, so the equation could be transformed to

$$E_{avg}(l) = \frac{1}{N} \sum_{i=1}^{N} \left| h_{li} \right| \left\| \beta_l \right\| \tag{7}$$

in which, $h_{li} = G(a_l, b_l, x_i)$, $E_{avg}(l) \in (0,+\infty)$, $E_{avg}(l) \to +\infty$ shows the sensitivity is greatest, otherwise the sensitivity is less. This leads to be another problem, for different real applications, it can not judge the degree of the sensitivity only by its value, because, in some applications, its value is very great, but in others, it is very little, there is no single criterion. So, another solution is given by the paper.
Given training samples (x_i, y_i), $i = 1,...,N$, the sensitivity based on each sample is $E(l,1), E(l,2),...., E(l,N)$. So, the sensitivity of the hidden node could be expressed by a point Se, $Se = (E(l,1), E(l,2),...., E(l,N))$, in space. In real applications, Se can not measure it directly, so, the Murkowski distance between Se and space original point O can be used when $q = 2$, it is the Euclidean distance.

$$d(Se,O) = \left(\sum_{i=1}^{N} \left| E(l,i) - O_i \right|^q \right)^{1/q} = \left(\sum_{i=1}^{N} \left| E(l,i) \right|^q \right)^{1/q} \tag{8}$$

$d(Se,O) \in (0,+\infty)$, so it is similar to $E_{avg}(l)$ and not consistence with people's cognitive habits, so it need to be transformed, its final result is

$$Sen(l) = e^{\frac{-\alpha}{d(Se,O)}} = e^{-\alpha \left\| \beta_l \right\|_q^{-1} \left(\sum_{i=1}^{N} \left| h_{li} \right|^q \right)^{-1/q}} \tag{9}$$

$Sen(l) \in (0,1)$, $Sen(l) \to 1$ shows the sensitivity is greater, otherwise it is less, which fully meets people's cognitive habits. Besides, it could measure the degree of sensitivity in different applications in one criterion. α is adjusting factor, which is used to manage the distribution of the sensitivity.

3.2 The output weight

The output weight of hidden nodes is the only parameter to be regulated during training, it is vital for ELM and its improved algorithms. After removing hidden nodes, the important task is to update the output weight, according to equation(5), matrix H and its inverse matrix H^{-1} should be obtained first. Set the sth hidden node is the removing hidden node, so the sth row and column of the matrix H is transformed to the Nth row and nth column, and the result matrix H_1' is further changed to

$$H_1' = \begin{bmatrix} H_1 & u \\ g & k \end{bmatrix} \tag{10}$$

in which, $u = [G(a_s,b_s,x_1)\cdots G(a_s,b_s,x_{s-1})\, G(a_s,b_s,x_{s+1})\cdots G(a_s,b_s,x_N)]^T$ is (N-1)×1 vector, $k = G(a_k,b_k,x_k)$ is a constant, $g = [G(a_1,b_1,x_s)\cdots G(a_{s-1},b_{s-1},x_s)\, G(a_{s+1},b_{s+1},x_s)\cdots G(a_M,b_M,x_s)]^T$ is 1×(N-1) vector. Referencing the literature[9], the inverse matrix of H_1' can be deployed as

$$H_1'^{-1} = \begin{bmatrix} (H_1 - ukg)^{-1} & -H_1^{-1}u\left(k - gH_1^{-1}u\right)^{-1} \\ -kg\left(H_1 - kug\right)^{-1} & \left(k - gH_1^{-1}u\right)^{-1} \end{bmatrix} \tag{11}$$

set $G = (H_1 - ukg)^{-1}$, so

$$H_1 = G^{-1} + kug \tag{12}$$

Referencing the literature[10], the inverse matrix of H_1 is

$$H_1^{-1} = G - \frac{GkugG}{1 + gGu} \tag{13}$$

So, output matrix H_1 and its inverse matrix H_1^{-1} is known, the output weight is,

$$\beta_1 = \left(H_1^T H_1\right)^{-1} H_1 Y_1 \tag{14}$$

where, $Y_1 = [y_1, \ldots, y_{s-1}, y_{s+1}, \ldots, y_N]^T$ 。

3.3 Description of the improved ELM

After detail analysis, the description of the improved ELM is

S1: Based on the specific problems, initial the threshold parameters ε and ε_{se}, and then it is to establish the SLFNs with enough hidden nodes by ELM, the number of hidden nodes M is less than the size M of training samples, M<N. Finally, according to the equation(5), the output weight β and output matrix H of hidden nodes is obtained.

S2: For each input training sample, the sensitivity $E(s,i)$ of each hidden node is obtained by equation(6) and then the sensitivity $Sen(l)$ can be computed by equation(9).

S3: For current inputting sample x_{N+1}, the output result y'_{N+1} of SLFNs could be obtained, for the given threshold ε, if the error of real output result y'_{N+1} and expected output result y_{N+1} meets $SSE(y_{N+1} - y'_{N+1}) < \varepsilon$, it could remove some hidden node with less sensitivity, go to Step S4, otherwise, Ready for the prediction or classification based on the next new sample.

S4: For the given threshold ε_{se}, if the sensitivity of the sth hidden node meets $Sen(s) < \varepsilon_{se}$, s<M, this hidden node could be removed.

S5: After removing hidden nodes, update the output weight by equation(14), and it continues to analyze if there is hidden nodes with less sensitivity, go to Step S3.

4. Tests

The performance of IPS-ELM is tested on the benchmark problems, inlcuding two regression problems (Auto-MPG and Abalone) and two classification problems (Image segment and DNA), SA-ELM[4] and ImSAP-ELM[3] are selected to be compared with IPS-ELM.

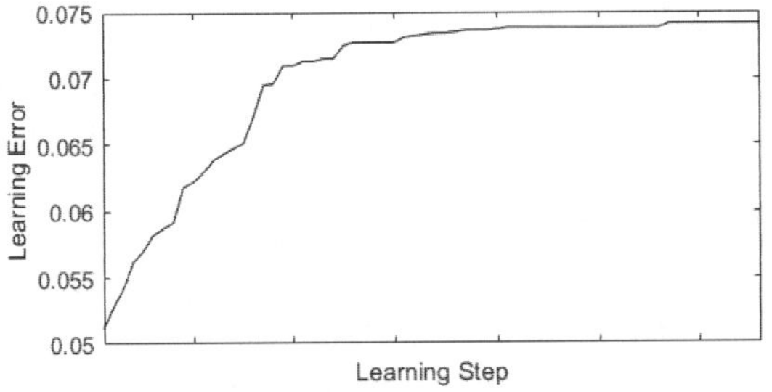

Figure 1. the variation curve of learning error.

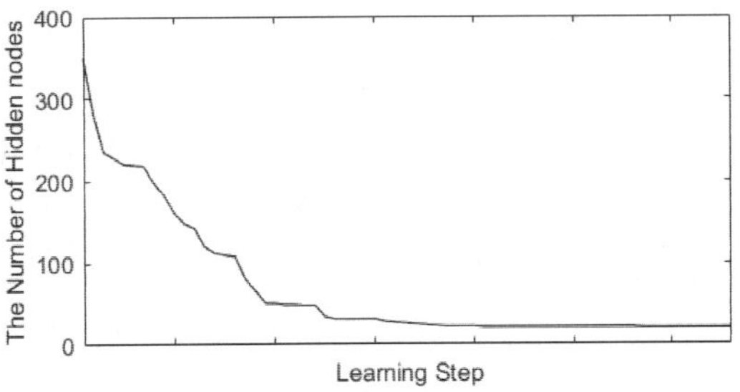

Figure 2. the variation curve of the number of hidden nodes.

Figure 1 and Figure 2 show the variation of learning error and the number of hidden nodes during the training respectively on Auto-MPG. From figures, the learning error is larger and larger with the number of hidden nodes decreasing, and the learning error finally reaches 0.073, it is fully acceptable.

Table 1. Comparison between IPS-ELM and other algorithms on regression problems

Algorithm	Auto-MPG			Abalone		
	Number of hidden node	Training time /s	error	Number of hidden node	Training time/s	error
IPS-ELM	21	1.6195	0.0008	12	0.1931	0.0007
SA-ELM	21	1.6287	0.0009	14	0.2161	0.0006
ImSAP-EM	23	2.0853	0.0021	20	0.2836	0.0012
ELM	52	1.9837	0.0027	25	0.0187	0.0037

Table1 is the testing results of the two regression problems, the finally error of the four algorithms is a little difference, but they are fully acceptable. The max error is 0.0037 of ELM on Abalone problem, and the minimum error is 0.0006 of SA-ELM on Abalone problem. So, compared with ELM, the accuracy of IPS-ELM is improved by 70% and 81% respectively. Compared with the improved ELM, it is improved by 74% and 50% respectively. The number of hidden nodes is smaller, can reach 21, so the SLFNs is more simple, learning time is shorter and can reach 1.6s and 0.19s.

Table 2. Comparison between IPS-ELM and other algorithms on classification problems

Algorithm	Image segment			DNA		
	Number of hidden node	Training time /s	error	Number of hidden node	Training time/s	error
IPS-ELM	49	0.1583	0.00322	156	0.4182	0.0051
SA-ELM	52	0.1975	0.00348	172	0.4837	0.0051
ImSAP-ELM	70	5.2183	0.00329	201	4.9235	0.0047
ELM	200	1.4	0.78	347	8.79	1.25

Table 2 is the testing results of the two classification, finally error (standard error) of the four algorithms is a little difference, in some applications, the accuracy of ELM is not acceptable, whose max error reaches 1.25. The classification accuracy of IPS-ELM is fully acceptable, the max is only 0.0051, equals with SA-ELM and only a little bigger than ImSAP-ELM by 0.0004, which is negligible. The number of hidden nodes by IPS-ELM is smaller, it is more simple, learning time is shorter and could reaches 0.15s and 0.41s respectively.

5. Conclusion

After deeply analyzing ELM and related algorithms, there are some weak points in their performances, such as the learning error, the number of hidden nodes and so on, so, based on these, a novel algorithm is proposed, it modified the analyzing method of the sensitivity of hidden nodes, which fully meets people's cognitive habits, based on the sensitivity, it removes the hidden nodes with less sensitivity on the promise of meeting the accuracy. After performance tests, IPS-ELM is better and acceptable.

6. ACKNOWLEDGEMENTS

This research is partially supported by the Science and Technology Funds of Dalian under Grant No. 2015A11GX010 and by Natural Science Foundation of Liaoning Province of China under Grant No.20170520196.

7. REFERENCE

[1] Huang G, Zhu Q, Siew C. Extreme learning machine: Theory and applications[J]. Neurocomputing. 2006, 70(1): 489-501.

[2] Haipeng C, Xiaohang T, Yu W, et al. Short-Term Parking Space Prediction Based on Wavelet-ELM Neural Networks[J]. Journal of Jilin University(Science Edition). 2017, 55(2): 388-392.

[3] DU Zhan-long, Xiao-Min L, Lei-Ping X, et al. Improved sensitivity-analysis based pruning extreme learning machine[J]. Control and Decision. 2016, 31(2): 249-255.

[4] Fan-Jun L, Hong-Gui H, Jun-Fei Q. Pruning algorithm for extreme learning machine based on sensitivity analysis[J]. Control and Decision. 2014, 29(6): 1003-1008.

[5] Li G, Liu M, Dong M. A new online learning algorithm for structure-adjustable extreme learning machine[J]. Computers & Mathematics with Applications. 2010, 60(3): 377-389.

[6] Alencar A S C, Rocha Neto A R, Gomes J P P. A new pruning method for extreme learning machines via genetic algorithms[J]. Applied Soft Computing. 2016, 44: 101-107.

[7] Yin J. A variable-structure online sequential extreme learning machine for time-varying system prediction[J]. Neurocomputing. 2017, 261(Supplement C): 115-125.

[8] Shao Z, Er M J. An online sequential learning algorithm for regularized Extreme Learning Machine[J]. Neurocomputing. 2016, 173(2016): 778-788.

[9] Peking U. Advanced Algebra (Second Edition)[M]. Beijing: Higher Education Press, 1987.

[10] Wenyu S, Yaxiang Y. Optimization theory and methods[M]. Beijing: Science Press, 1997.

A Novel Multi-Camera Calibration Method based on Flat Refractive Geometry

S Huang[1], M C Feng[1,3], T X Zheng[1], F Li[2], J Q Wang[1] and L F Xiao[1]

[1]School of Advanced Manufacturing Engineering, Chongqing University of Posts and Telecommunications, Chongqing 400065, china
[2]School of Automation, Chongqing University of Posts and Telecommunications, Chongqing 400065, china
[3]fengmc@cqupt.edu.cn

Abstract. Multi-camera calibration plays an important role in many field. In the paper, we present a novel multi-camera calibration method based on flat refractive geometry. All cameras can acquire calibration images of transparent glass calibration board (TGCB) at the same time. The application of TGCB leads to refractive phenomenon which can generate calibration error. The theory of flat refractive geometry is employed to eliminate the error. The new method can solve the refractive phenomenon of TGCB. Moreover, the bundle adjustment method is used to minimize the reprojection error and obtain optimized calibration results. Finally, the four-cameras calibration results of real data show that the mean value and standard deviation of the reprojection error of our method are 4.3411e-05 and 0.4553 pixel, respectively. The experimental results show that the proposed method is accurate and reliable.

1. Introduction

Multiple cameras calibration system is a key problem in computer vision research, and it has been become more and more attention recently. As camera calibration is indispensable step from two-dimensional (2D) images to extract three-dimensional (3D) space information, so this technology has been widely used in 3D space object measurement, virtual reality, 3D reconstruction, and other fields.
At present, many kinds of calibration methods have been proposed in multi-camera system. For example, Pribanić used an orthogonally contacted 3-axis wand for multi-calibration [1,2]. Jia proposed an approach to multiple cameras calibrate based on LCD template [3]. Svoboda developed a useful multi-camera self-calibration method [4]. Hui presents a series of scanning calibration methods using 3D patterns [5,6]. Sun developed a global calibration of multiple cameras based on sphere targets, which can be easily manipulated and observed from different directions [7]. However, each method has its own limitation of the accurate and feasible.
To solve above problem, a new multi-camera calibration method based on flat refractive geometry is proposed in the paper. Besides, calibration targets usually use planar boards in metric calibration. In the paper, the TGCB is used with the help of calibration question. Compared with Traditional calibration methods, the new multi-camera calibration method has a higher accuracy and feasible and it can also effectively improve the quality of 3D reconstruction.

2. Camera model

2.1. camera projection model

The camera imaging model is the basis of the camera calibration. In the camera calibration, first of all, we need to know the image pixel coordinate system, image physical coordinate system, the camera coordinate system and the world coordinate system position relationship. uOv is established in the image coordinate system, the coordinates of each point on the image (u,v) on behalf of each pixel in the image of the number of columns and rows. As (u,v) corresponds to the value of the image at this point is the grey value, so (u,v) called pixel coordinates, which (u_0,v_0) is the pixel coordinates of the center point of the image.
Figure 1 show the $O_c - X_cY_cZ_c$ represents the camera coordinate system, where O_c is the optical center of the camera. X_c , Y_c are parallel to the x axes and y axes of the image coordinate system, respectively. The choice of the world coordinate system is generally based on the actual situation to carry on the reasonable choice, the $O_W - X_WY_WZ_W$ represents the world coordinate system. The world coordinate system and the camera coordinate system transformation relations as follows:

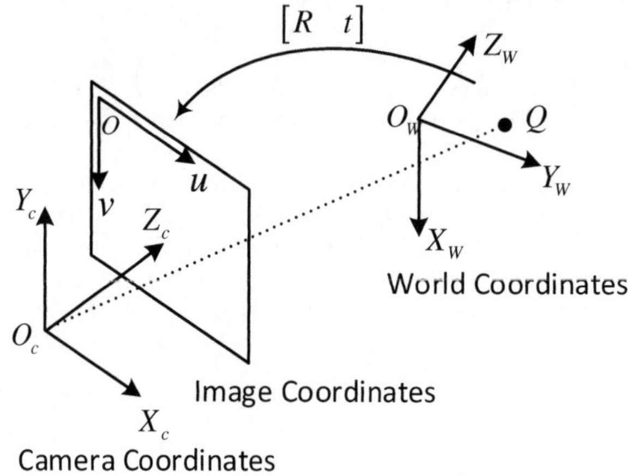

Figure 1. The relationship between the world coordinate system and the camera coordinate system.

$$\begin{bmatrix} X_c \\ Y_c \\ Z_c \\ 1 \end{bmatrix} = \begin{bmatrix} R & t \\ 0 & 1 \end{bmatrix} \begin{bmatrix} X_W \\ Y_W \\ Z_W \\ 1 \end{bmatrix} \tag{1}$$

where, the vector t is the coordinate value of the origin of the world coordinate system in the camera coordinate system. R is the rotation matrix of the world coordinate system to the camera coordinate system.
The world coordinate system 3D point to image coordinate system 2D point is described by:

$$\begin{bmatrix} u \\ v \\ 1 \end{bmatrix} = K \begin{bmatrix} R & t \\ 0 & 1 \end{bmatrix} \begin{bmatrix} X_W \\ Y_W \\ Z_W \\ 1 \end{bmatrix}, K = \begin{bmatrix} f_u & 0 & u_0 \\ 0 & f_v & v_0 \\ 0 & 0 & 1 \end{bmatrix} \tag{2}$$

where, the vector f_u and f_v are the focal length in the image axes u and v .

2.2. transparent glass calibration board (TGCB)
The TGCB can provide the 3D coordinates of feature points, which are used as corner points. Several corner points are uniformly distributed on the TGCB.

In this paper, the TGCB shown in the Figure 2 is a standard planar template. Moreover, we can calibrate cameras with Zhang's method [8]. This TGCB is simple to manufacture and easy to carry, therefore, it could be effectively applied to site calibration projects.

Figure 2. Transparent glass calibration board (TGCB).

2.3. flat refractive geometry

The flat refractive geometry is introduced to solve the refractive problem of the TGCB. The TGCB is used as calibration targets, which contains the refractive surface and non-refractive surface. Figure 3 shows the flat refractive projection model. The black and white grids pattern were printed on the side λ_1 of the TGCB, and the camera of this side can directly observe black and white grids pattern. The camera on the other side can observe black and white grids pattern through the glass. Where, P_1 is the intersection of refractive surface and imaging ray. P_2 is the spatial 3D point coordinates, which assumed to be known. d_0 indicates the distance from the camera to the TGCB, and d_1 represents the thickness of the TGCB. r_0 denotes the direction vectors of light path in the air, and r_1 denotes the direction vectors of refractive light path in the glass. The camera's extrinsic parameter $[R\,t]$ is transformation between the world coordinate system O_W and the camera coordinate system C .

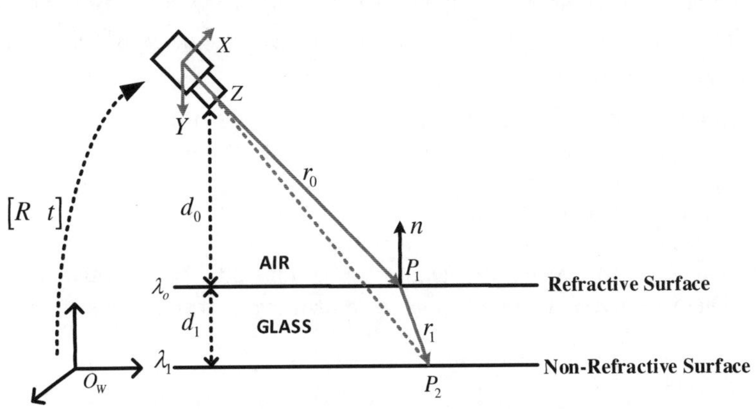

Figure 3. Flat refractive projection model.

The transformed from the world coordinate system to the camera coordinate system is presented as:

$$p_\gamma = KP_1 = K[[R\,t]P_2 + d_1 \frac{r_1}{Rnr_1}] \tag{3}$$

where, K represents the camera's intrinsic parameters. When the ray r_0 of the camera don't pass through refractive surface of the TGCB, and directly intersect at the 3D point P_2 of the TGCB.

3. Proposed method

3.1. multi-camera calibration based on refraction model

In order to obtain accurate calibration parameters of the multi-camera system, we present a novel method based on flat refractive geometry. The multi-camera system consists of four cameras which distribute on both sides of the calibration target. Each camera captures an image of the calibration target from different positions and orientations. The proposed method can accurately calculate the parameters of multi-camera system by using Zhang's method [8] and the flat refractive geometry.

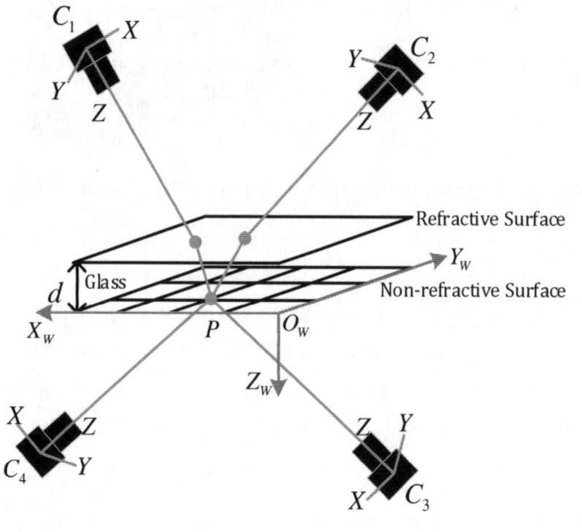

Figure 4. The calibration scene of multi-camera.

The specific forms of the multi-camera refractive projective model are defined in this section. Figure 4 depicts the calibration scene of multi-camera, which distributes on both sides of the calibration target, where one side of the camera has a refractive phenomenon.

Assume that the thickness of the TGCB is accurately measured. The non-reflective surface of the TGCB is defined as the world coordinate system, and the world coordinates of 3D point is denoted by $P=(X\ Y\ Z)^T$. When the point P is projected to four cameras at the same time, the C_1 , C_2 camera receive projection rays through refractive surface, while the C_3 , C_4 camera will receive the projective rays without going through refractive surface. The projection rays of the camera without through the refractive surface by using the traditional calibration method, and through the refractive surface by using the refractive projection model for calibration.

3.2. Nonlinear optimization

The camera parameters are optimized by using the bundle adjustment method to minimize the reprojection error. The objective function for bundle adjustment is expressed as:

$$\min_{D,K,R,T,\gamma} \omega = \sum_{k=1}^{L}\sum_{j=1}^{n}\sum_{i=1}^{m} w\|\tilde{p}-\tilde{p}'(D,K,R,t)\|^2 + (1-w)\|\tilde{p}_\gamma-\tilde{p}_\gamma'(D,K,R,t,\gamma)\|^2 \tag{4}$$

where, $\tilde{p}$, $\tilde{p}'$ are actual pixels point and measured point without refraction. $\tilde{p}_\gamma$, $\tilde{p}_\gamma'$ are actual pixels point and measured point with refraction. Projection parameters contain distortion parameter D, intrinsic parameter K , rotation matrix R , translation vector t and refractive index γ . w is the refractive weight, which indicates whether the ray is refracted. L represents camera number, n indicates the number of captured images, and m represents the number of 3D point. The equation (4) is solved by the bundle adjustment method based on the Levenberg-Marquardt algorithm. Then, we get the optimal value of the multi-camera system obtained by minimizing the reprojection error.

4. Experiments

In this experiment, an actual four-camera system is applied to calculate the extrinsic parameters of the system which is similar to Figure 4. The four-camera system is displayed in Figure 5. Six groups of images are captured and one of them is shown in Figure 6. The calibration results of four cameras are relatively complex, so the intrinsic and extrinsic parameters of the cameras are not shown.

Figure 5. The setup for four cameras system for calibration experiments.

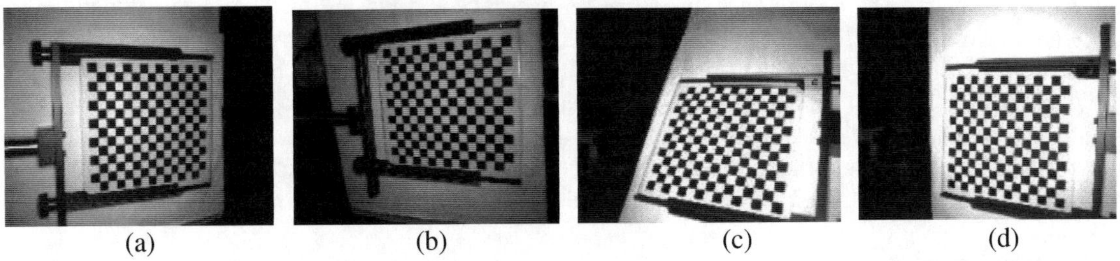

Figure 6. A group of sample images. (a) Calibration image of the first camera. (b) Calibration image of the second camera. (c) Calibration image of the third camera. (d) Calibration image of the fourth camera.

In order to prove the accuracy and reliability of the proposed method, the frequency distribution of reprojection error of the four cameras is counted. From Figure 7-8 we can see that the mean and standard deviation of reprojection errors change from -0.3378 and 2.954 pixel to 4.3411e-05 and 0.4553 pixel. The calibration accuracy of the four-camera system is significantly improved by refinement. Therefore, the real data of four-camera system inspect the accuracy and reliability of the proposed method.

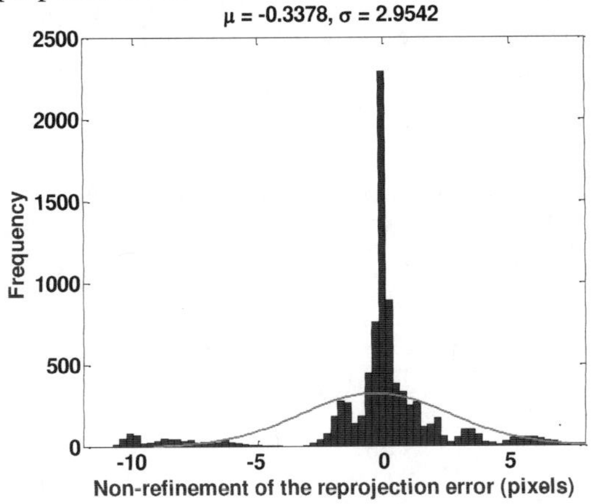

Figure 7. Frequency distribution histogram of the reprojection error without refinement (four camera).

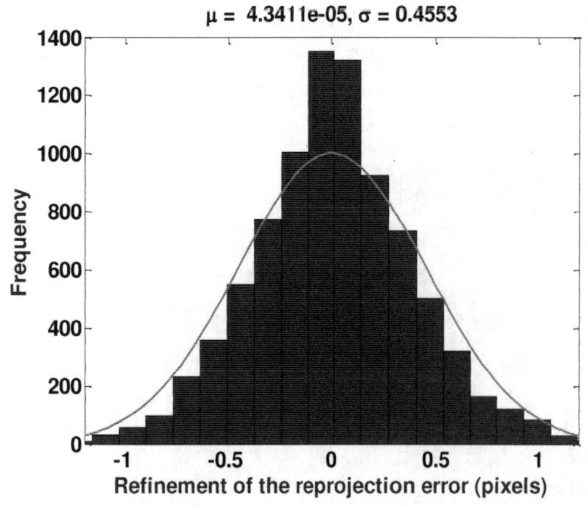

Figure 8. Frequency distribution histogram of the reprojection error with refinement (four camera).

5. Conclusion

In this paper, we developed a novel calibration method for multi-camera system based on flat refractive geometry. This method achieves a higher accuracy and feasibility. The TGCB is applied as calibration target and all cameras can acquire calibration images at the same time from different orientations. Due to the refractive phenomenon of TGCB, the proposed calibration method also solves the phenomenon that can be used for one-shot measurement application. The four-camera calibration results of real data show that the mean value and standard deviation of the reprojection error of our method are 4.3411e-05 and 0.4553 pixel, respectively. The result of the experiments also confirmed the accuracy and feasibility of the proposed method.

6. Acknowledgments

This works is jointly supported by Scientific and Technological Innovation of Chongqing (Grant No. CYS17233), National Natural Science Foundation of China (NSFC) (Grant No. 51505054), Chongqing Municipal Education Commission (Grant No. KJ1500405), Chongqing Science and Technology Commission (Grant No. cstc2015zdcy-ztzx60002, cstc2016jcyjA0538, cstc2015zdcy-ztzx30001).

7. References

[1] Pribanić T, Sturm P and Peharec S 2009 Wand-based calibration of 3D kinematic system *IET Computer Vision* **3**(3) pp 124-129

[2] Pribanić T, Sturm P and Cifrek M 2007 Calibration of 3D kinematic systems using orthogonality constraints *Machine Vision and Applications* **18**(6) pp 367-381

[3] Jia Q Q, Wang B X and Zhao B H 2009 Global calibration of multi-CCD line-structured-light measurement system based on LCD target *Optics & Precision Engineering* **17**(12) pp 2990-2996

[4] Svoboda T, Martinec D and Pajdla T 2005 A convenient multicamera self-calibration for virtual environments *PRESENCE: teleoperators and virtual environments* **14**(4) pp 407-422

[5] Hui B, Wen G and Zhao Z 2012 Line-scan camera calibration in close-range photogrammetry *Optical Engineering* **51**(5) pp 1-12

[6] Hui B, Wen G and Zhang P 2013 A novel line scan camera calibration technique with an auxiliary frame camera *IEEE Transactions on Instrumentation and Measurement* **62**(9) pp 2567-2575

[7] Sun J, He H, and Zeng D 2016 Global calibration of multiple cameras based on sphere targets *Sensors* **16**(1) p 77

[8] Zhang Z 2000 A flexible new technique for camera calibration *IEEE Trans. Pattern Anal. Mach. Intell* **22** pp 1330-1334

IOP Conf. Series: Materials Science and Engineering **320** (2018) 012017 doi:10.1088/1757-899X/320/1/012017

The application and development of artificial intelligence in smart clothing

XIONG Wei

College of Textile Garment, Southwest University, Chongqing 400700, China

Abstract: This paper mainly introduces the application of artificial intelligence in intelligent clothing. Starting from the development trend of artificial intelligence, analysis the prospects for development in smart clothing with artificial intelligence. Summarize the design key of artificial intelligence in smart clothing. Analysis the feasibility of artificial intelligence in smart clothing.

1. Introduction

Artificial intelligence is a combination of computer science, control theory, information theory, neurophysiology, psychology, linguistics, philosophy and so on the many kinds of mutual penetration and development of the subjects of a cross discipline. [1] Google's Alpha Go against the would champion at Go Lee se-dol and KeJie become the new world champion in May 2017. The development of Alpha Go makes us excited about development of artificial intelligence. A professor at the university of Massachusetts institute of technology in United States Swinton has publicly "Artificial intelligence is to study how to make computer to do some work only humans can do before". The Alpha Go as one of artificial intelligence is the imitation board game between the human activity. [2] In the early years, a number of well-known fashion designers have applied high technology to fashion design, and then smart clothes gradually appeared in the public view. Smart clothing refers to the simulated life system, which can not only perceive the external environment, but also the change of internal state, and the clothing can respond to this change in real time. [3] In the intelligent life environment, artificial intelligence and smart clothing can adapt to people's various wear needs.

2. The Development Prospects of Artificial Intelligence in The Smart Clothes

2.1 The necessity of intelligence transformation traditional garment industry

The concept of "industrial 4.0" began to emerge in the new round of revolution of science and technology which is made "smart" as the leading the fourth industrial revolution. Artificial intelligence will become the focus of the development direction and the virtual world will fuse with the real world in the era of industrial 4.0. The future manufacturing will achieve higher engineering efficiency use shorter time to market and achieve production flexibility. [4] From the point of view of the world development, the development of artificial intelligence to the world economy is more and more fast in the arrival of industrial 4.0. According to the Accenture global twelve in developed economics. Artificial intelligence will help them to speed up the economic growth. Based on the traditional theory, capital and labor are the two elements of mass production of boosting the economy. [5] The traditional pattern of industrial development is made up of capital, labour and total factor productivity. Capital and labor are variable elements that determine economic development. Adding artificial intelligence to the new industrial development model makes it another variable element associated with capital and labour development, which will be more efficient in promoting economic development.

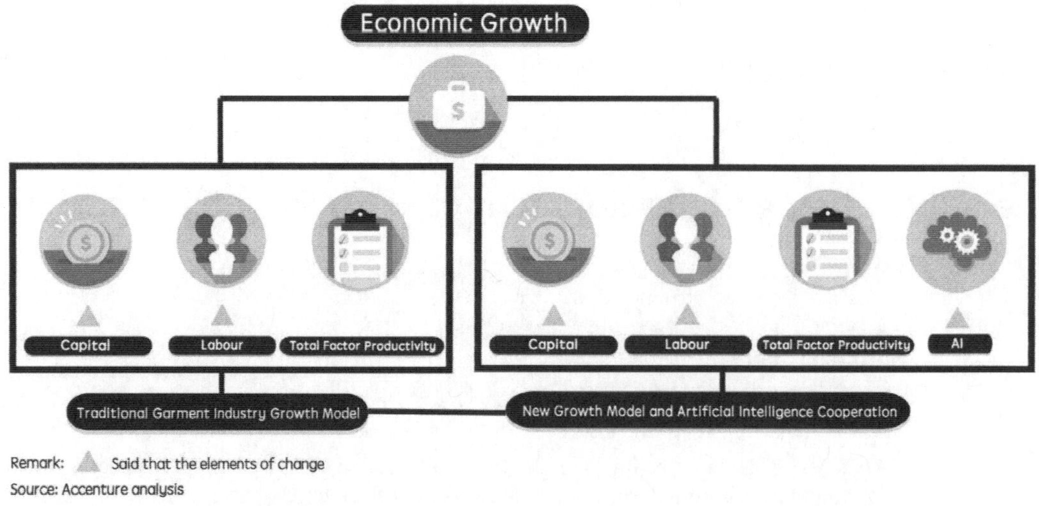

Figure 1. AI as a new factor's influence on the economic growth of the apparel industry.

China, as a major labour force, has abundant labour resources. As a basic consumer goods industry, garment industry is highest degree of marketization in China, one of the strongest industry international competitiveness. The rapid development of China's economy and the expansion of the national dress of consumer demand, is the fundamental motivation for the development of China's garment industry. At the same time, high-quality labor resources, good investment environment and to form a complete set of complete and well versed in upstream and downstream industry chain, for the development of China's garment industry provides an important conditions and powerful guarantee. For the apparel industry of intelligent transformation provides a good bedding.

2.2 Efforts of artificial intelligence in the costume design

Some clothing need the multi-discipline frontier technology. It is primarily through two kinds of methods to implement the intelligent of the clothing. One is using intelligent clothing materials, including shape memory materials, phase change materials, colour materials and stimulus-response hydrogels, etc. The second category is the microelectronics technology and information technology in People's Daily wear clothing, including the application of conductive material, flexible sensor, unlimited communication technology and power supply, etc. [6] At present, some fashion designers began to design with artificial intelligence's assistance. With the development of science and technology in the future, artificial intelligence can be inserted in the clothing to become human's closest aides.

2.2.1 Artificial intelligence and cognitive dress. In early 2016, IBM will work with Marchesa in the theme of " Manusx Machina: Fashion in an Age of Technology" the Met Gala. They built the first cognitive dress for it. IBM's artificial intelligence Watson has been involved in the design of the dress. It not only provides the inspiration for the design of the dress, its technology is woven into the fabric of the dress. Dress in 150 flower embroidered on a built-in LED lights. LED by Watson Tone Analyzer API real-time monitoring label content changes which with # Met Gala# and# Cognitive Dress# in Twitter. It can analysis the emotions to change colours in real time. First of all, Watson study science report accurately choose the right material. Watson analysis massive gallery choose the five most popular colour for LED lights. Then designers started to design production of the dress. When the model walked down the aisle, Watson began to read fan mood real-time feedback change colour. In addition, Watson has a good ability to learn, through expert training and optimization of constant progress. Australia's fashion designer Jason Grech using Watson "visual identity" to analysis of the past decade fashion data and social information in time. In order to analysis and forecast trend, so as to help him find a new way to deal with material, color and texture. Watson also help designers build image and fashion image to match each other. It makes designers convenient from line, curve and inspiration in the edges and corners. Watson cross-border cooperation with designers tell us that artificial intelligence can make a designer more diverse and creative.

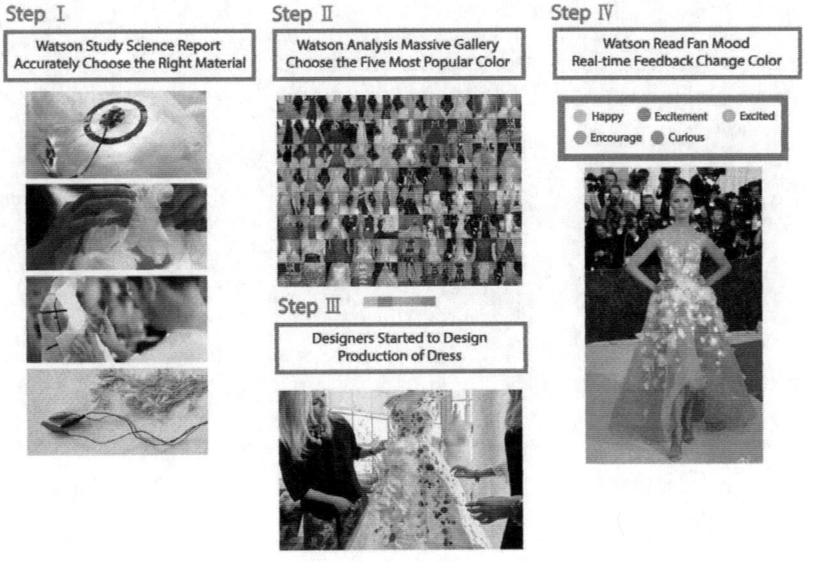

Figure 2. The process of Watson with fashion designer to design the dress.

2.2.2 Related to intelligent technology for AI clothing design inspiration. Face recognition refers to the input face image or video, determine whether there is a face. If there is further human face, and then gives the location of each face, size and the location of the main facial organs. And based on these information, further extract each face is the identity of the characteristics, and comparing with the known face the library face, so as to identity of each face. [7] In the design of smart clothing, this technology can be combined with artificial intelligence. It can make the garment itself to face recognition, and to identify the face of the corresponding effects such as colour, fabric shrinkage, etc. Or directly to store the human face information in the memory in order to extract information when it needed. Such as the artificial intelligence to help its memory across the person's identity information. Through the wearable device for human-computer interaction. "wearable devices" broadly understood that can wear in the body of the electronic equipment, or into a dress or similar dress a portable electronics equipment. From a practical sense, "wearable device" is not just a dress or a similar dress to electronic equipment, but also be able to use data interaction. The cloud interaction and software support to realize the special function can be worn in or worn on the body of technology products. Fore example, Intel has started using wearable device for people to interact with technology. Grammy awards ceremony in 2016, Lady Gaga with Intel's Digital Skin, Interactive Video, Robotics, Interactive Holograms in collaboration with the performance. Through take based on Intel Curie platform development of wearable devices to control the show which with the human body in the big screen effect. In the design of intelligent clothing, it can join wearable devices to interact better with artificial intelligence or wearable device using the electronic, circuit into clothing, in order to achieve the functional requirements. In the design of wearable devices can be micron or nanometre embedded, make the garment of sensors and drive with wireless connection between the surrounding environment. At the same time, the micro or nanometre electronic components such as small size can better invisibility within the clothing, to make the dress more fashionable appearance. In this way, there is a visual gap between wearable devices and ordinary clothes which can basically be invisible and minimize. [8] The emergence of new flexible energy storage elements has greatly improved the comfort of smart clothing and wearable devices, which has brought infinite possibilities to the design of artificial intelligent clothing.

Figure 3. The dynamic hologram of in Lady Gaga' s Face

2.2.3 AI clothing design key point. Traditional clothing is mainly composed of ordinary fabric cutting, the ain function is to clothing, beautiful and fashionable. And smart clothing is mainly composed of intelligent materials, such as optical fabric, memory fabric, etc. Smart clothing need to consider electronic components configuration issues, both to ensure that the intelligent and want to wear comfortable. [9] we also need to consider in the design of clothing and artificial intelligence interactive equipment, the availability of intelligence clothing material, and form a complete set of services for intelligent clothing combined with artificial intelligence. The focus of the artificial intelligence clothing design can be summarized as the following several aspects. *Clothing Wearing Comfort.* Smart clothing is intelligent technology and clothing integrated high-tech products. It is a combination of electronic information technology, sensor technology, textile science and cutting-edge technology in the field of material science or related. Due to the combination of a variety of new frontier technology, inevitably will use new type of smart materials and wearable devices to improve the intelligent level of clothing. Smart clothing while considering the functional. It need to consider when fabric or wearable devices with the comfort and safety of human body skin contact with the human body. Consider the question of whether or not to accord with human body engineering, etc. *The user need.* Smart clothing as a function of clothing, the main purpose is to serve a particular user. It as a carrier of wearable devices. It is necessary to serve different needs and withstand the subjective choice of buyer's market. Due to the different of the users living environment, social class, education environment, lead to their different acceptance of new intelligent products. Designers need to pay attention to the release of new technologies and integrate with the development of science and technology to better adapt to the needs of users. This needs constantly to evaluate the users cognitive, the experience of new intelligent products and the acceptance of the artificial intelligence emotion in the design. These meet the user's physiological and psychological needs.

Interactions with The Equipment of Artificial Intelligence and Cognition. Because smart clothing need induction for interacting with artificial intelligence. Artificial intelligence need to people's emotion. Designers should be considered in the design of clothing or wearable devices in use when the operation is convenient. Smart clothing is for a specific customer service. So designers should consider the user's point of view in a simple operation that artificial intelligence to complete more efficient instruction.

Provide Supporting Software and Services. Smart clothing is based on soft ware systems and the internet platform. It with the intelligence of the combination with cloud computing and big data. Fashion designer need to consider with specific software test procedure and the combination of digital media to artificial intelligence clothing which collection of data for storage and management. In order to better will be the result of the user needs to feedback to the user.

3 The feasibility of artificial intelligence in the smart clothes and analysis the difficulties in design

3.1 the feasibility of artificial intelligence clothing

From the choice of the market and the future trend of the society, smart clothing to meet the needs of people and cater to the development of the times. In such an era of intelligent fast rhythm, the emergence of smart clothing is a necessity. Artificial intelligence industry investment increased year by year; The relevant policies of artificial intelligence has risen to national level; Artificial intelligence will become the commanding heights of the enterprise strategy layout; Artificial intelligence in some professional task performance is often better than humans. In such an era of intelligent fast rhythm, the emergence of smart clothing is a necessity.

3.2 Artificial intelligence clothing promotion difficulty

The use of artificial intelligence in the smart clothes also has the obvious disadvantages. The smart clothing is in the theoretical test phase in the current. If we want a wind range of promotion, we need to overcome a lot of difficulties. The future need to overcome the difficulty has the following several.

3.2.1 Design and Technical Difficulties. In intelligent garment design need to consider all aspects. Many creative ideas are not achievable because of fabric constraints or technological backwardness. To apply theory to practice, still need to conquer all kinds of technical difficulties. Such as smart clothing clean energy storage components in water whether to go bad and wearable equipment maintenance, etc. We need to development the science and technology to solve these problems.

3.2.2 The Prise Problem. Smart clothing at the time of design need to use special intelligent fabrics or intelligent wearable devices. And these fabrics and equipment cost much more expensive than ordinary fabrics. A smart clothing prices than the price of the same style of clothing is much higher. So, the price will be a big factor restricting the development of smart clothing. Such as smart clothing clean energy

3.2.3 Security Issues. Smart clothing is shared with the Internet resources. It also makes many kinds of intelligent clothing involves the user's private information, location information and other sensitive information. It gave hacker a lot of opportunities to steal the information. This also reflected the development of intelligent clothing is a double-edged sword. On the one hand, it has facilitated people's lives. On the other hand, it also provides a more criminals. In the future, if we want to promotion the intelligent clothing, designer should thoughtful and be better to improve the safety.

3.2.4 Difficult Points to Promote in The Market. Consumers in the market have different levels of consumption level and consumption view. Many consumers are sitting on the fence about new smart products. This makes it difficult to promote many new smart products. On the one hand, smart clothing is more expensive than ordinary clothing. It is also in the experimental phase of its use. This makes most consumers give up buying for these reasons. On the other hand, due to the imperfection of technology, many intelligent products boring poor user experience to consumers. These are the inevitable problems in the process of smart clothing promotion.

4 Conclusion

Today, intelligent has become a trend in the development of the era. The development of science and technology, the development and application of new technology materials provide guarantee for the future development of smart clothing. Through the constant progress and development of science and technology, we can predict the future is the development of textile and clothing industry has a broad prospect. Smart clothing will become an important indicator of garment industry. And the development of artificial intelligence can also make smart clothing have more limitless possibilities. From the development prospects of artificial intelligence in the clothing design. The rapid development of science and technology, the transformation of clothing industry is inevitable, combining science and technology development is also necessary. Designers and Watson successful cooperation let us see the clothing design and artificial intelligence cooperation in the future. As long as we have confidence in smart clothing. It will assist in the development of artificial intelligence industry have their new era.

5 References

[1] Zhu F and Li 2017.06 *Technology and application of artificial intelligence Alpha Go analysis* (China CIO News) p 107

[2] Wu H L 2009.02 *First met artificial intelligence* (Science and Technology Innovation Herald) p196

[3] Zou F Y 2008.01 *Design and Exploitation of Intelligent Clothing* (Art & Design)

[4] *Industrial age 4.0- artificial intelligence* 2017.05 (Smart Factory) pp 33-34

[5] Mark P and Qiu J 2017.04 *Accenture: artificial intelligence can help China's economic growth* (Robot Industry) pp 80-91

[6] Tang S L P and Stylios G K 2006 *An overview of smart technologies for clothing design and engineering* (International Journal of Clothing Science & Technology) pp 108-128

[7] Li X L 2014.03 Artificial intelligence theory and its application in face recognition (Science & Technology Vision) p 148

[8] Huang T L 2017.05 *Clothing structure and craft research in wearable devices* (Beauty & Times) pp 102-106

[9] Qin J Y and Cao S 2017.06 Interaction design of smart clothing based on quantified self under the concept of green IT design for sustainability (Packaging Engineering) pp 1~6

IOP Conf. Series: Materials Science and Engineering **320** (2018) 012018 doi:10.1088/1757-899X/320/1/012018

Longitudinal Control for *Mengshi Autonomous Vehicle* via Cloud Model

H B Gao[1], X Y Zhang[2], D Y Li[3] and Y C Liu[3]

1 State Key Laboratory of Automotive Safety and Energy, Tsinghua University Beijing, China
2 Information Technology Center, Tsinghua University, Beijing, China
3 The Institute of Electronic System Engineering, Beijing, China
E-mail: ghb48@mail.tsinghua.edu.cn

Abstract. Dynamic robustness and stability control is a requirement for self-driving of autonomous vehicle. Longitudinal control method of autonomous is a key technique which has drawn the attention of industry and academe. In this paper, we present a longitudinal control algorithm based on cloud model for Mengshi autonomous vehicle to ensure the dynamic stability and tracking performance of Mengshi autonomous vehicle. An experiments is applied to test the implementation of the longitudinal control algorithm. Empirical results show that if the longitudinal control algorithm based Gauss cloud model are applied to calculate the acceleration, and the vehicles drive at different speeds, a stable longitudinal control effect is achieved.

1. Introduction

The basic functions of automated longitudinal vehicle control are keeping the vehicle a safe distance behind another vehicle, maintaining a relatively constant speed with the least brake use and applying the brake as fast as possible in emergency situations. The challenges handled in the design of the longitudinal control system include nonlinear vehicle dynamics, string-stable operation with very small inter-vehicle spacing, operation at all speeds from a complete stop to high-speed cruising, and the execution of longitudinal split and join maneuvers in the presence of communication constraints. Lijuan Wu and his colleagues present a platoon control algorithm based on fuzzy logic to achieve the longitudinal control of a platoon of automotive vehicles [1]. S. Sheikholeslam and Charles A. Desoer propose control laws in the event of loss of communication between the lead vehicle and the other vehicles in the platoon, which solves the problem of longitudinal control of a platoon of automotive vehicles on a straight lane of a highway [2]. P. Zheng, M. McDonald and W. Lang et al, by use of fuzzy logic, act to control of vehicle in car following [3][4]. G. Qiang, H. Shuliang and D. Chao describe a car following controller based on fuzzy inference [5]. S. Kumarawadu and T. Lee survey a longitudinal controller based on neural network [6]. A. Ghaffari and his colleagues by use of Emotional Learning Fuzzy Inference System (ELFIS) approach, design a longitudinal controller for car following [7]. H. M. Kim et.al designs a combined brake/throttle fuzzy controller that uses a neural system to learn the fuzzy rules to control the velocity and the distance between cars in single-lane platoons [8]. L. Cai, AB. Rad and W.Lok Chan present a neuro-fuzzy controller for intelligent cruise control of semiautonomous vehicles. This approach aims at regulating the speed of the controlled vehicle in order to maintain constant time headway with respect to the vehicle in front [9]. H.I Lee and M. Tomizuka, with the method of adaptive vehicle traction force control, achieve the robust longitudinal control of vehicles in intelligent vehicle highway systems [10]. Ya-Fu Peng et. al proposes a robust intelligent backstepping control (RIBC) scheme for the car-following control of a platoon of automated vehicles using a recurrent cerebellar model articulation controller (RCMAC) via the H∞ control technique [11]. AFerrara, P. Pisu, L. Nouveliere and S. Mammar propose longitudinal controllers that rely on the generation of "second order" sliding regimes [12][13]. X.Yun Lu and Karl J. Hedrick consider the theoretical part for longitudinal control of merging maneuver for Automated Highway System [14]. L.Daxue, S. Zhenping, Z. Hanqing, Li Zheng and H. Hangen design a feedforwd controller for longitudinal control of autonomous ground vehicle based on the analysis of the vehicle force in off-road Environment [15]. M. Tai and M. Tomizuka use traction control and brake control solving the robust longitudinal velocity tracking of vehicles [16].

The current study aims to improve the accuracy, robustness, and adaptability to various road conditions of the vehicle control algorithm. First, the convergence of vehicles toward trajectory tracking errors is investigated from the perspective of nonlinear system stability, which is the premise of vehicle tracking trajectory. Subsequently, the robustness and control algorithm that can adapt to the environment is also considered. Finally, the function of vehicle motion control is expanded, which enables vehicles to complete automatic overtaking task, adaptive cruise task, automatic parking task, flowing into traffic task and so on.

Research on vehicle control cited above, some researches only focused on lateral tracking control, some researches only focused on longitudinal tracking control, without considering learning from people driving. When autonomous driving tasks increase in complexity, the control systems cited earlier are unable to adapt to complex tasks. In addition, the control system should be able to guarantee stability. The main contributions of our study are as follows. (1) A new uncertainty control system according to the Gauss cloud model (GCM) and cloud reasoning is illustrated. (2) The new model learning from human driving, dealing with complex tasks. (3) The longitudinal control algorithm for autonomous driving vehicles are constructed, with reference to human driving experience.

This paper is organized as follows. Section 1 presents the longitudinal control of autonomous vehicle. Section 2 presents the GCM, the GCM algorithm, and cloud reasoning, including a preconditioned Gauss cloud generator (PGCG), a post-conditioned Gauss cloud generator (PCGCG). Section 3 describes the longitudinal control rule and algorithm for autonomous vehicle. Section 4 provides the results of the experiment and analysis performed using the longitudinal control algorithm based on cloud model. Finally, the results of experiment are illustrated in Section 5.

2. Model and problem formulation

2.1. Gauss Cloud Model

The Gauss distribution (GD) is one of the most important distributions in probability theory, in which the general characteristics of random variables are represented as means of the mean and variance of two numbers. As a fuzzy membership function, the bell-shaped membership function is mostly used in sets, which is typically expressed through the analytical expressions of $m(x) = \exp\{-(x-a)^2/2b^2$. This study presents a cloud model based on the GD, called the Gauss cloud model (GCM), which is defined as follows [17][18].

DEFINITION 1: U is expressed in a precise numerical quantitative domain. *C(Ex, En, He)* is a qualitative concept on U. If the value of x ($x \in U$) is a random realization of the qualitative concepts of C, then the "Expectation" of the GD $x \sim N(Ex, En'^2)$ is denoted as Ex, and its "Variance" is denoted as En'^2. Meanwhile, the "Expectation" of the GD $En' \sim N(En, He^2)$ is denoted as En, and its "Variance" is denoted as He^2. En' is the full form of the GD $En' \sim N(En, He^2)$ and is a random realization[19]. The certainty degree of x in C is satisfied via $m(x) = \exp\{-(x-Ex)^2/2(En')^2\}$. The distribution of x in the domain of U is called a Gauss cloud (GC) [20]. The GC algorithm is presented in table 1 [20].

Table 1. The GC algorithm.

Input: Three figures (Ex, En, He) and the number of cloud drops n.
Output: A sample set that represents concept extension and its certainty (x_i, m_i), i

(1) To generate a Gauss random $En' \sim N(En, He^2)$
(2) To generate a Gauss random $x \sim N(Ex, En'^2)$
(3) To calculate the certainty: $m(x) = \exp\{-(x - Ex)^2/2(En')^2\}$
(4) Repeat (1)-(4) until the number of cloud drops is n.

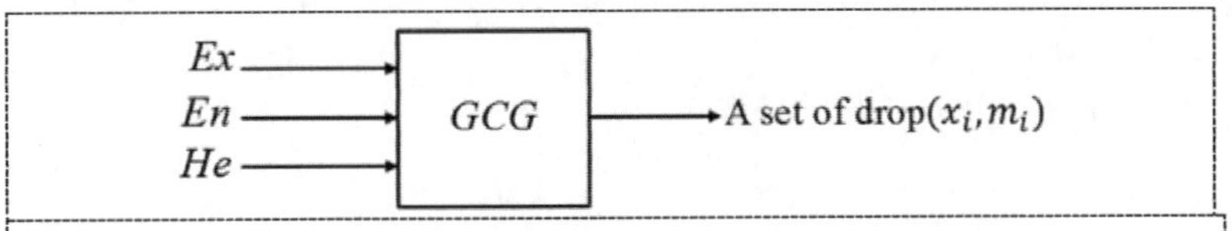

Figure 1. The GCG.

The algorithm causes distribution drops, called cloud distribution (CD). The algorithm of GCM can be obtained through a cloud generator (CG), which forms a forward Gauss cloud generator (GCG), as shown in figure 1. The Gauss random number generation method is the foundation of the whole algorithm. It generates uniform random numbers in [0, 1] and uses them to calculate the Gauss random number. Random number sequences are determined through the uniform random function of a seed. The method of using uniform random numbers to generate a Gauss random number is described in detail in Ref. [21]. GC distribution (GCD) is different from the GD because the GCD algorithm uses the Gauss random number twice, in which one random number is the basis of another random number. Among these,

(1) When *He* = 0, the algorithm generates a precise value of *En* and the value of *x* is transformed into a GD.

(2) When *He* = 0, *En* = 0, the value of *x* of the algorithm generation is an exact value of *Ex*, and $m \equiv 1$.

From (1) and (2), certainty can be concluded as a special case of uncertainty, and the GD is a special case of the GCD.

For a qualitative concept of a steering angle of positive and negative 40°, given that *Ex* = 80°, *En* = 1, and *He* = 0.1, 1000 cloud drops are generated. The distribution of drops and its certainty degree of *C* (*x*, *m*) are shown in figure 2.

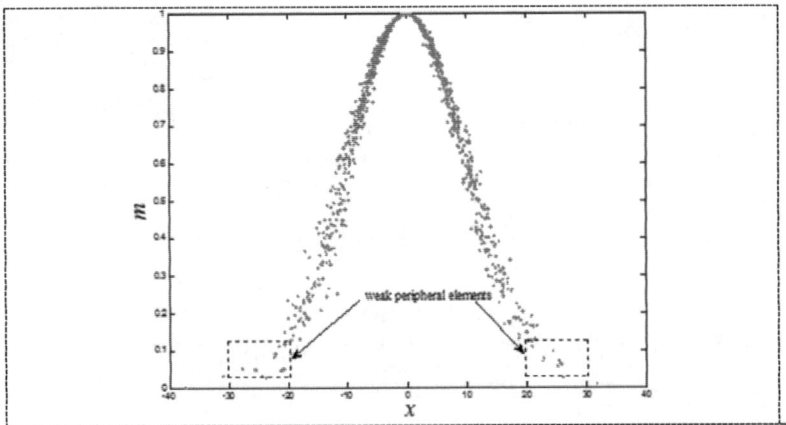

Figure 2. The distribution of 1000 drops.

2.2. Cloud Reasoning

2.2.1. Preconditioned Gauss cloud generators. Knowledge forms a concept and its relationship with communicating and abstracting. The relationship among concepts form certain rules, from which rules library and rules generator can be established through knowledge reasoning based on GC. Rules include preconditioned and post-conditioned rules. Preconditioned rules include one or several rules, whereas post-conditioned rules express the results and specific control actions generated by the preconditioned rules. In the control field, "perception–action" can establish the rule library based on the relationship among concepts, thereby realizing control of uncertainty.A preconditioned Gauss cloud generators (PGCG) and a post-conditioned Gauss cloud generators (PCGCG) are composed of the GCG, which is defined as follows.

DEFINITION 2: Assume the following rule:

If A, then B,

where A corresponds to concepts C_1 in universal sets U_1, and B corresponds to concepts $C2$ in universal sets U_2. a is a specific value in universal sets U_1, where the GCG generates a specific value of a based on the concept C_1 of the certainty degree of m distribution, and $m \in [0,1]$, which is called a PGCG [22].

2.2.2. Post-conditioned Gauss cloud generators.

DEFINITION 3: Assume the following rule:

If A, then B,

where A corresponds to concepts C_1 in universal sets U_1, and B corresponds to concepts C_2 in universal sets U_2. The certainty degree of m belongs to [0, 1]. The GCG generates the certainty degree of m drop distribution, which is satisfied by applying concepts C_2 in universal sets U_2, called PCGCG [23] [24].

3. Longitudinal control based on cloud model

3.1. Data Analysis

As shown in figure 3, the longitudinal velocity control model is established according to the cloud model and cloud reasoning. The input of the longitudinal velocity control model is the pedal opening, the velocity of the vehicle, the acceleration of the vehicle and the expected velocity of the vehicle. The output is the acceleration required for the vehicle to reach the expected velocity [25].

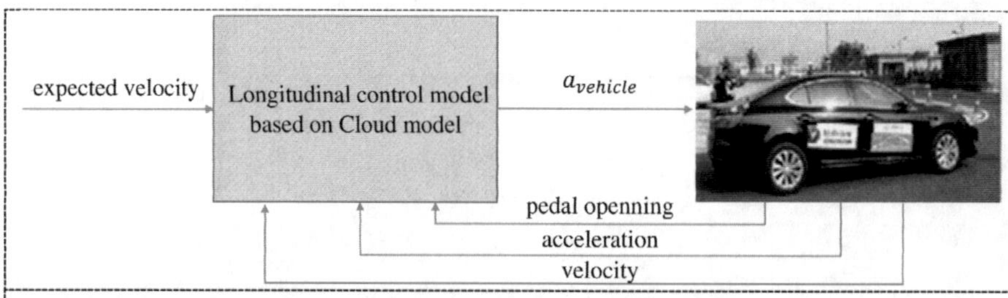

Figure 3. Longitudinal control data model based on cloud model.

Table 2. The Numerical characteristics of cloud model of *a*.

v	$v\ (E_x, E_n, H_e)$
Positive greater	(9.8,1.1,0.18)
Positive less	(4.9,1,0.19)
Zero	(0,1,0.01)
Negative less	(−4.7,1,0.03)
Negative greater	(−9.8,1.2,0.03)

Table 3. The Numerical characteristics of cloud model of *a*.

a	$a\ (E_x, E_n, H_e)$
Positive greater	(19,2.6,0.045)

Positive less	(9,2.1,0.02)
Zero	(0,1.9,0.007)
Negative less	(− 9,2,0.02)
Negative greater	(19,2.8,0.05)

The real car test in the city of Beijing lasted a month, test 4 hours per day, the data recorded frequency of 1Hz. The detection data including the pedal opening, the velocity of the vehicle, the acceleration of the vehicle and the expected velocity of the vehicle. The data volume of 400,000 above. The domain of the quantitative attribute is divided into the concept of the cloud model. Table 2 and table 3 show the three digital characteristics (*Ex*, *En*, *He*) of vehicle velocity, vehicle acceleration [26].

3.2. Longitudinal Control Rules and Algorithm

The longitudinal control of the autonomous vehicle is single input and single output controller, the input of the cloud controller is the difference between expected velocity and actual velocity of autonomous vehicle Δv. The output is the acceleration a. The variable a can be described using five qualitative concepts, namely, 'positive greater', 'positive less', 'near-zero', 'negative less', and 'negative greater'. The input and output variables define the five qualitative concepts and construct a corresponding cloud regulation generator.

Assume the following rule: If E, then F, where E is the PGCG that generates the drop distribution (a, m) with a specific value of a and a certainty degree of m. E is replaced by the cloud model digital feature of the velocity obtained by the experiment, then a represents the velocity of the autonomous driving vehicle, m represents the degree of determination of the velocity, the degree of determination of the velocity is related to the acceleration. F is the PCGCG that generates the drop distribution (b, m) of the cloud with a specific value of b and a certainty degree of m. F is replaced by the numerical model of the cloud model of the acceleration obtained by the experiment, then m is the degree of determination of this acceleration, b is the acceleration required for the autonomous driving vehicle to reach the expected velocity, which is called longitudinal control model based on cloud model, which is shown in figure 4 [25].

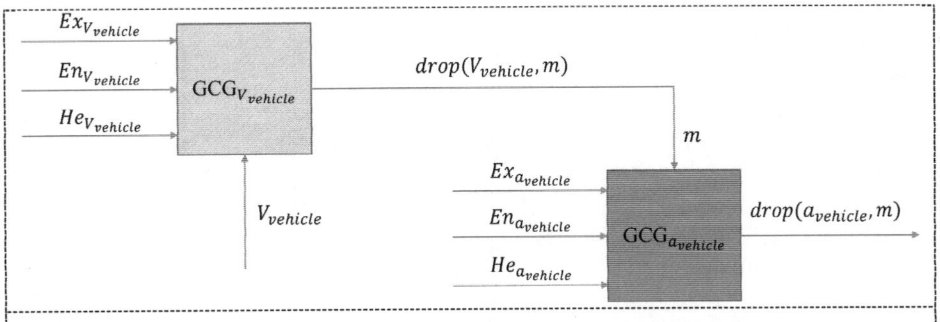

Figure 4. The longitudinal control algorithm chart. The longitudinal control algorithm is presented in.

The longitudinal control algorithm implies an uncertainty transfer in the conceptual reasoning process. In the universal sets U_1 of the PGCG, the distribution of the certainty degree of m belongs to the specific value of $v_{vehicle}$, whereas the certainty degree of m is the input of the PCGCG that generates the drop distribution ($a_{vehicle}$, m) of the cloud specific value of $a_{vehicle}$ and the certainty degree of m. The processing of the certainty value of $v_{vehicle}$ to the certainty value of $a_{vehicle}$ is uncertain [23][25][26].

4. Experiment result and analysis

4.1. Experiment Setup

IOP Conf. Series: Materials Science and Engineering **320** (2018) 012018 doi:10.1088/1757-899X/320/1/012018

4.1.1. Hardware architecture of an autonomous vehicle system. The on-board sensor configuration of an autonomous vehicle comprises a radar sensor, a vision sensor, and a positioning sensor. The radar sensor consists of a 32 lines laser radar on the vehicle, a forward SICK single laser radar, a forward four lines laser radar, and a backward millimeter wave radar. The vision sensor comprises three front-facing cameras, two rear-facing cameras, and two lateral cameras set in both rear-view mirrors. The positioning sensor consists of the Global Positioning System (GPS) and an inertial measurement unit (IMU), as is shown in figure 5. All types of sensors are mainly applied to sense the surroundings of the vehicle for real-time acquisition of its location, posture, velocity, and time.

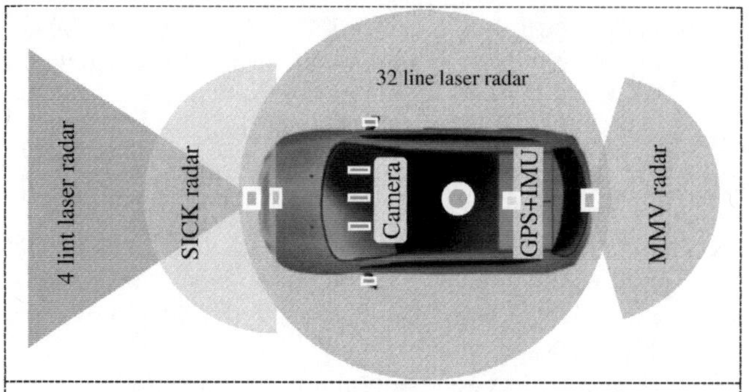

Figure 5. Experiment sensor configuration.

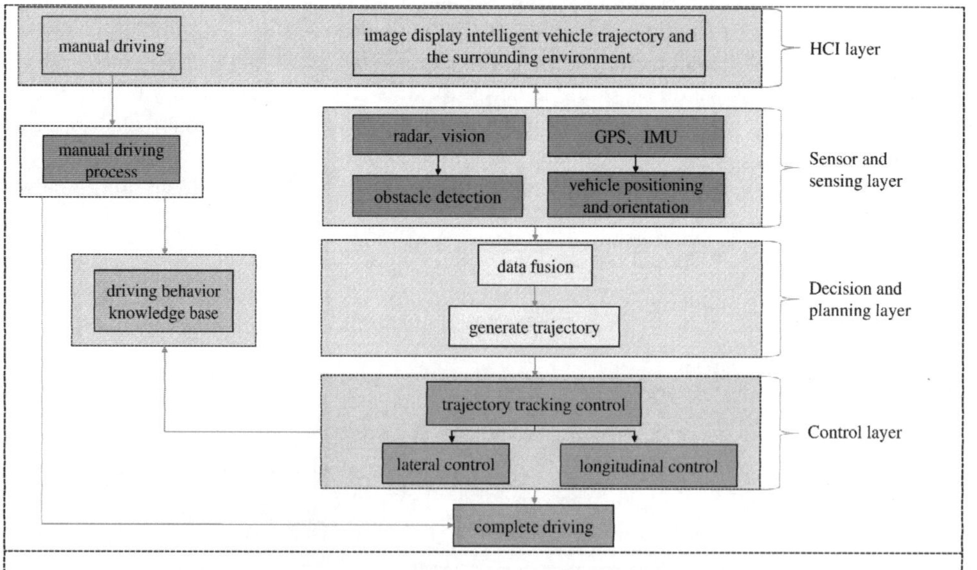

Figure 6. Software architecture of an autonomous vehicle system.

4.1.2. Software architecture of an autonomous vehicle system. The software architecture of autonomous vehicle systems is shown in figure 6. This architecture comprises a human computer interaction (HCI) layer, a sensor and sensing layer, a planning and decision layer, and a control layer.
HCI layer: This layer receives the touch commands and emergency braking instructions of the driver.
Sensor and sensing layer: This layer consists of a radar sensor, a vision sensor, a GPS sensor, and an IMU sensor. The sensor focuses on completing the collection of sensor data. The sensing layer focuses on sensor data analysis, road edge identification, obstacle detection, traffic sign detection, and body state estimation, which complete data fusion.
Decision and planning layer: This layer focuses on path planning and decision, which determine the driving pattern of an autonomous vehicle by analyzing environment data and vehicle data from the

sensory module. The outputs is Acceleration and the increment of steering angle. Those commands output to control layer.

Control layer: This layer directly outputs the control order to the accelerator, as well as the braking and steering controller, of the vehicle. It also receives human instructions and performs acceleration/deceleration and steering operations.

4.1.3. Experimental environment. The Beijing-Tianjin Expressway, which spans the Taihu Toll Station and the Dongli Toll Station, covers 121 km of shuttle distance. Rain is moderate rain in Tianjin, with a small amount of water on the ground. The weather is rainy in the Tianjin section of the Beijing-Tianjin Expressway. When the sun occasionally shines, the weather remains sunny until reaching Beijing, where it is cloudy. The temperature outside the vehicle is 32°C, and that on the road is 40°C. Visibility is over 200 m.

4.2. Experiment Result and Analysis

When the autonomous vehicle proceeds, the instant velocity is obtained using GPS, and the acceleration is obtained using inertial measurement unit (IMU).

4.2.1. Speed analysis. As shown in figure 7, the speed curve of Tianjin to Taihu Toll. As show in figure 8, the speed curve of Taihu Toll to Tianjin, the *x* axis represents the time of data, and the data recording interval is 200ms; The *y* axis indicates the speed, unit: km/s, Tianjin to Taihu station, the mileage is 86km, the average speed is 87.26km/h and program set speed is 90 km/h; Taihu to Tianjin station, the mileage is 86km, the average speed is 94.1km/h and program set speed is 100 km/h; The situation of speed changing reflects the driving pattern. In the experiment, within the first 30 kilometers, it used 90km of the vehicle search driving model. During the experiment, the speed of the Beijing-Tianjin high-speed on his car speed fast, the autonomous driving vehicle without.

Slow speed vehicles and speed maintained in 90km/h. Due to road construction, implement the artificial intervention, the speed is set to 50km/h, and driving with cars, the speed affected by the front car is kept at 60-100km/h. The figure shows a data speed exceeding 150km/h, which is an invalid speed value.

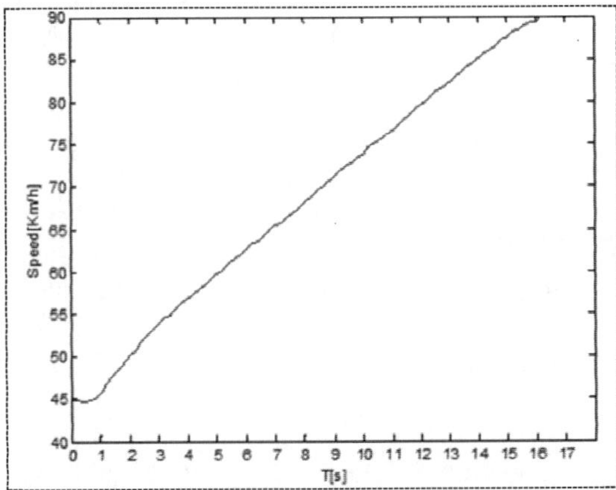

Figure 7. Speed curve of Tianjin to Taihu Toll.

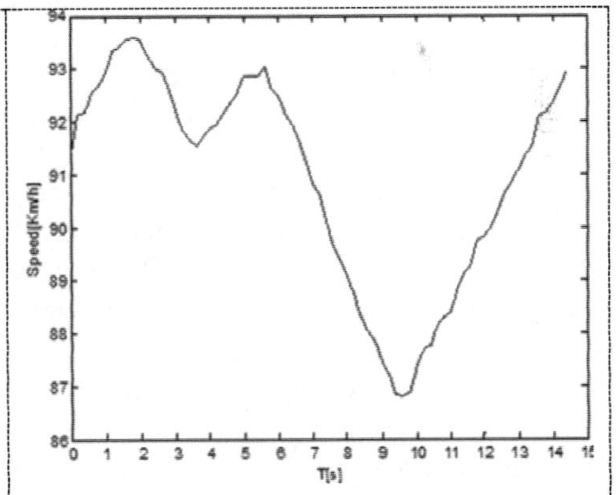

Figure 8. Speed curve of Taihu to Toll Tianjin.

4.2.2. Acceleration analysis. As shown in figure 9, the acceleration curve of Tianjin to Taihu Toll. As shown in figure 10, the acceleration curve of Taihu Toll to Tianjin, the *x* axis represents the time of data, and the data recording interval is 200ms; The *y* axis indicates the acceleration, unit: km/s^2. In the experiment, the acceleration value of the autonomous vehicle during acceleration and deceleration is maintained at -2m/s^2~1m/s^2. Korea k. Yi use the linear optimal control theory to design the upper control, considering the need of vehicle ride comfort, the output of the upper controller is saturation limited, the vehicle longitudinal acceleration is limited to -2m/s^2~1m/s^2 range. In rare cases, the acceleration occurs more than 1m/s^2, causing a rapid acceleration and the passenger's subjective feeling is better.

The experimental results show that the longitudinal control based on cloud model can achieve good control effect on speed and acceleration, the speed and acceleration have strong stability and small fluctuation. The phase plane of velocity and acceleration shows a certain diversity, reflecting the phenomenon of real experiment. The longitudinal control based on cloud model compared with other control methods, such as PID control and fuzzy logic control, the results of the longitudinal velocity control based on cloud model is not the only constant, it can reflect the uncertainty of longitudinal velocity control.

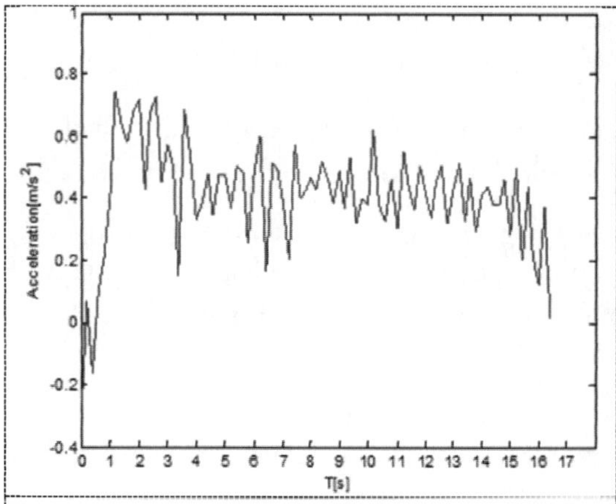

Figure 9. Acceleration curve of Tianjin to Taihu Toll.

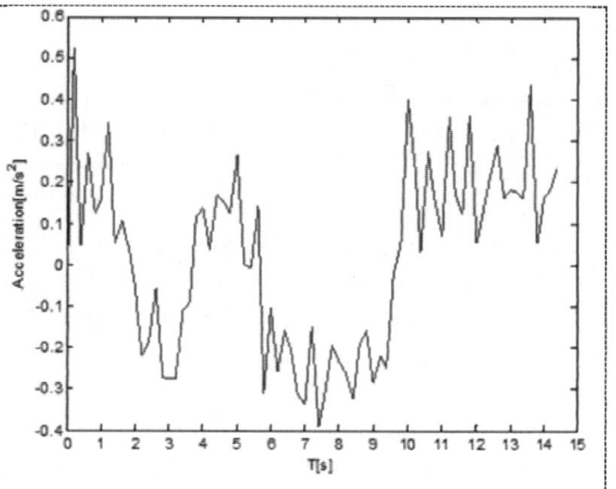

Figure 10. Acceleration curve of Taihu to Toll Tianjin.

5. Conclusion

This paper proposes the application of cloud model and cloud reasoning to the longitudinal control of autonomous vehicles. Because the cloud model use expected value, entropy and hyper entropy to characterize the qualitative concept, so it can integrate the fuzziness and randomness together in qualitative and quantitative conversion to overcome the inherent defects of the membership function in fuzzy set theory. The uncertainty of the control of autonomous driving vehicle has the fuzziness and randomness, cannot be expressed by the precise mathematical model, and the control of the driver pedal operation is realized by using the cloud inference, expressed the randomness of the speed control. In this paper, we classify the speed and acceleration of the cloud model, classify it according to experience, no certification, therefore, this will be the follow-up research work, by the introduction of cloud transformation, the data in the same concept cluster together, the data between different concepts are classified, fully reflect the actual distribution characteristics of the data.

6. Acknowledgments

This work was supported by National Natural Science Foundation of China under Grant No.61035004, No.61273213, No.61300006, No.61305055, No.90920305, No.61203366, No.91420202, No.61571045, No.61372148, the National High Technology Research and Development Program ("863"Program) of China under Grant No.2015AA015401, and the National High Technology Research and Development Program ("973"Program) of China under Grant No. 2016YFB0100903, and the Junior Fellowships for Advanced Innovation Think-tank Program of China Association for Science and Technology under Grant No.DXB-ZKQN-2017-035, and the Beijing Municipal Science and Technology Commission special major under Grant No. D171100005017002.

7. References

[1] Wu L, Chen X and Ling Y 2006 The Application of Fuzzy Control Algorithm in the Longitudinal Control System for a Platoon of Vehicles *Proc. of the 6th World Congress on Intelligent Control and Automation* (Dalian, China) pp 21-23
[2] Sheikholeslam S and Desoer C A 1993 Longitudinal Control of a Platoon of Vehicles with no Communication of Lead Vehicle Information: A System Level Study *IEEE Transactions on Vehicular Technolog* **42**(4) pp 546-554
[3] Zheng P J and McDonald M 2005 Application of Fuzzy Systems in the Car-Following Behavior, Analysis, Lecture Notes in Computer Science *Fuzzy Systems and Knowledge Discovery* **3613** pp 782-791
[4] Lang W, Wang C and Chiang Y On the car-following model with fuzzy control *Proc. 2nd Nat. Conf Fuzzy Theory & Application(Fuzzy'94)* pp 380-385
[5] Qiang G, Shuliang H and Chao D 2008 The Modeling and Simulation of the Car-following Behavior Based on Fuzzy Inference *International Workshop on Modelling, Simulation and Optimization* (Hong Kong, China) pp 322-325
[6] Kumarawadu S and Lee T 2004 A Model Based Neuro control Approach for Car Following Collision Prevention *Proc. of the IEEE int. Conf. on Networking, Sensing and Control* (Taipei, Taiwan) vol 1 pp 152-157
[7] Demetriou K, Kapazoglou A, Bladenopoulos K and Tsaftaris A S 2012 MANFIS-based overtaking maneuver modeling and prediction of a Driver-Vehicle-Unit in real traffic flow. *Proc. of IEEE Int. Conf. on Vehicular Electronics and Safety* (Istanbul, Turkey) vol 28 pp 387-392
[8] Kim H M, Dickerson J and Kosko B 1996 Fuzzy throttle and brake control for platoons of smart cars *Fuzzy Sets Syst* **84** pp 209-234
[9] Cai L, Rad A B and Chan W L 2010 An Intelligent Longitudinal Controller for Application in Semiautonomous Vehicles *IEEE Transactions on Industrial Electronics* **57**(4) pp 1487-97
[10] Lee H I and Tomizuka M 2003 Adaptive Vehicle Traction Force Control for Intelligent Vehicle Highway Systems (IVHSs) *IEEE Transactions on Industrial Electronics* **50**(1) pp 37-47
[11] Peng Y F, Hsu C F, Lin C M and Lee T T 2006 Robust Intelligent Backstepping Longitudinal Control of Vehicle Platoons with $H\infty$ Tracking Performance *Int. Conf. on Systems, Man, and Cybernetics* (Taipei, Taiwan) pp 4648-53
[12] Ferrara A and Pisu P 2004 Minimum Sensor Second-Order Sliding Mode Longitudinal Control of Passenger Vehicles *IEEE Transactions on Intelligent Transportation Systems* **5**(1) pp 20-32
[13] Nouveliere L and Mammar S 2003 Experimental Vehicle Longitudinal Control. using Second Order Sliding Modes *Proc. Of the American Control Conf. Denver* (Colorado, USA) vol 6 pp 4705-10
[14] Yun L X and Hedrick J K 2000 Longitudinal Control Algorithm for Automated Vehicle Merging *International Journal of Control* **76** pp 193-202
[15] Da X L, Zhen P S, Han Q Z, Zheng L and Han G H 2008 Longitudinal Controller Design For Autonomous Ground Vehicle in Off-road Environment *Proc. of the 27th Chinese Control Conf.* (Kunming, Yunnan, China) pp 687-690
[16] Tai M, Tomizuka M 2000 Robust Longitudinal Velocity Tracking of Vehicles Using Traction and Brake Control *6th Int. Workshop on Advanced Motion Control. Proc.* (Nagoya, Japan) pp 305-310
[17] Li D R, Wang S L and Li D Y 2015 Cloud Model *Spatial Data Mining*
[18] Li D Y, Liu Y C and Du Y 2004 Artificial Intelligence with Uncertainty *Journal of Software* **15**(11) pp 1538-94
[19] Cao B H, Li D Y and Qin K 2010 An Uncertain Control Framework of Cloud Model *Int. Conf. on Rough Set and Knowledge Technology* (Beijing, China) pp 618-625
[20] Li D Y, Liu Y C and Gan W Y 2009 A new cognitive model: Cloud model *International Journal of Intelligent Systems* **24**(3) pp 357-375
[21] Gao H X 1995 *Statistical calculation* (Beijing: Peking University Press)

[22] Li D Y and Du Y 2005 Artificial Intelligence with Uncertainay (Beijing: National Defence Industry Press) pp 137-185
[23] Gao H B, Jiang J, Zhang L Y, Liu C and Li D Y 2014 Cloud Model: Detect Unsupervised Communities in Social tagging Network *Int. conf. on information science and cloud computing* (Guangzhou, China) pp 317-323
[24] Gao H B and Zhang X Y 2016 Research of intelligent vehicle variable granularity evaluation based on cloud model *Acta Electronica Sinica* **44**(2) pp 365-373.
[25] Du Y 2000 *Research and Applications of Association Rules in Data Mining* (PLA University of Science and Technology) pp 34-45.
[26] Li D Y, Du Y and Yin G D 2000 Commonsense Knowledge Modeling *Proc. of 16th World Computer Congress* (Beijing, China) pp 34-45